Nervous Conditions

SUNY SERIES,
STUDIES IN THE LONG NINETEENTH CENTURY

Pamela K. Gilbert, editor

SUNY SERIES IN
SCIENCE, TECHNOLOGY, AND SOCIETY

Sal Restivo and Jennifer Croissant, editors

NERVOUS CONDITIONS

Science and the Body Politic in Early Industrial Britain

ELIZABETH GREEN MUSSELMAN

STATE UNIVERSITY OF NEW YORK PRESS

Published by
STATE UNIVERSITY OF NEW YORK PRESS
ALBANY

© 2006 State University of New York

For information, address
State University of New York Press,
194 Washington Avenue, Suite 305, Albany, NY 12210-2384

Production, Laurie Searl
Marketing, Anne M. Valentine

Library of Congress Cataloging-in-Publication Data

Green Musselman, Elizabeth, 1971–
    Nervous conditions : science and the body politic in early industrial
Britain / Elizabeth Green Musselman.
        p.   cm. — (SUNY series in science, technology, and society)
    (SUNY series, Studies in the Long Nineteenth Century)
    Includes bibliographical references and index.
    ISBN 0-7914-6679-5 (hardcover : alk. paper)
    1. Scientists—Mental health—Great Britain—History—19th century.
    2. Nervous system—Philosophy—History—19th century.
    3. Science—Philosophy—History—19th century.   I. Title.   II. Series.
    RC464.A1G74   2006
    616.8'001'9—dc22

2005014025

ISBN-13 978-0-7914-6679-7 (hardcover : alk. paper)

10   9   8   7   6   5   4   3   2   1

for
Jack and Liam

Written after Recovery from a Dangerous Illness

Lo! o'er the earth the kindling spirits pour
  The flames of life that bounteous Nature gives;
The limpid dew becomes the rosy flower,
  The insensate dust awakes, and moves, and lives.

All speaks of change, the renovated forms
  Of long-forgotten things rise again;
The light of suns, the breath of angry storms,
  The everlasting motions of the main.

These are but engines of the Eternal will,
  The One Intelligence, whose potent sway
Has ever acted, and is acting still,
  Whilst stars, and worlds, and systems all obey.

Without whose power, the whole of mortal things
  Were dull, inert, an unharmonious band,
Silent as are the harp's untuned strings
  Without the touches of the poet's hand.

A sacred spark created by his breath,
  The immortal mind of man his image bears;
A spirit living 'midst the forms of death,
  Oppress'd but not subdued by mortal cares. . . .

      —Humphry Davy, quoted in
  John Davy, *Memoirs of the Life of Sir Humphry Davy*

Contents

PREFACE   ix

PART I
Embodied Epistemology

ONE   The Nervous Man of Science   3

TWO   The Social Hierarchy of Subjectivity   30

PART II
The Nervous Conditions

THREE   Provincialism and Color Blindness   55

FOUR   Mental Governance and Hemiopsy   101

FIVE   Rational Faith and Hallucination   146

SIX   Conclusion   189

NOTES   199

BIBLIOGRAPHY   229

INDEX   267

Preface

Anyone who has transformed a doctoral dissertation into a book has incurred a long thank-you list. It is only appropriate to thank first those who served as midwives to the birthing process. For generously sharing their time, expertise, and encouragement with me, I thank my doctoral committee at Indiana University: Richard Sorrenson, Jim Capshew, Jeanne Peterson, and Nico Bertoloni-Meli. I have also enjoyed generous support at my professional home-away-from-home, the Department of History and Philosophy of Science at Cambridge University. In particular, I thank Simon Schaffer, Anne Secord, Jim Secord, Paul White, and Martin Kusch. Long conversations with each of them have improved my thinking immeasurably.

While a Mellon postdoctoral fellow in Oklahoma's Department of the History of Science, I benefited from a stimulating intellectual environment and the wonderfully rich history of science collections there. Karen and Larry Estes, Keri and Chris McReynolds, Gregg Mitman, Katherine Pandora, Sylvia Patterson, Kim Perez, and Kathleen Wellman made my stay in Norman enjoyable and productive.

Since then, I have had the fortune to teach at Southwestern University, where the support is even bigger than Texas. For putting such a bright face on my first years of full-time academic work and for encouraging me to keep working on this book, I am especially grateful to Eileen Cleere, Ed Kain, Sahar Shafqat, Kim Smith, Julie Thompson, and my unequaled department colleagues—Daniel Castro, Steve Davidson, Jan Dawson, Lisa Moses Leff, and Thom McClendon.

I owe a great debt to many other history of science colleagues whose insights, support, and kind (but incisive) criticism have made them true friends. Kathy Olesko nurtured my first interests in the history of science during my undergraduate years at Georgetown University. For their help with this project, I especially thank Will Ashworth, Brad Hume, Bruce Hunt, Iwan Rhys

Morus, Michael Reidy, Klaus Staubermann, Jenny Tannoch-Bland, Jennifer Tucker, and Norton Wise. Participants in the July 1998 Bellagio Center conference on "The Interpenetration of Science and Philosophy of Science," organized by David Hull, engaged me in informative conversations on the topics discussed in this book.

A number of institutions provided generous support for my research, including a three-year graduate fellowship from the National Science Foundation, a Mellon postdoctoral fellowship from the University of Oklahoma, and several grants from Indiana University and Southwestern University.

The following librarians made my archival research both productive and pleasant: Pat Fox and Cliff Farrington (Harry Ransom Humanities Research Center, University of Texas at Austin); Sandra Cumming and Mary Sampson (Royal Society of London); Godfrey Waller and Adam Perkins (Cambridge University Library); Elizabeth Quarmby (St. John's College, Cambridge); Patricia Methven (King's College London); Peter McNiven (John Rylands University Library of Manchester); Sylvia Patterson (University of Oklahoma); and the library staff at Indiana University; University of Massachusetts–Amherst; Mt. Holyoke College; Southwestern University; Harvard University's Houghton and Widener Libraries; the British Library; Trinity College, Cambridge; University College London; Edinburgh University; and the Wellcome Library for the History and Understanding of Medicine.

For permission to republish some of the material in this book, I thank the editors at *History of Science* and Berg Publishers. Much of chapter 3 also appeared in my paper, "Local Colour: John Dalton and the Politics of Colour Blindness," *History of Science* 38 (2000): 401-24. Portions of chapters 4 and 6 were previously published in "The Governor and the Telegraph: Mental Management in British Natural Philosophy," in *Bodies/Machines,* ed. Iwan Rhys Morus (Oxford and New York: Berg Publishers, 2002). For permission to quote manuscripts, I thank the Syndics of Cambridge University Library and the Particle Physics and Astronomy Research Council, the Harry Ransom Humanities Research Center at the University of Texas at Austin, King's College London, the Director and Librarian at the John Rylands University Library of Manchester, and the Master and Fellows of Trinity College Cambridge.

For making my first book publishing experience rewarding, I thank James Peltz, Laurie Searl, and Alan V. Hewat at State University of New York Press. Two anonymous reviewers commissioned by the press gave me extensive and constructive feedback.

Finally, I offer my heartfelt gratitude to my family. My parents, Tom and

Julie Green, and sister, Mary, have always believed in me and the value of my work. My parents-in-law, John and Mary Musselman, have encouraged me with their faith in the importance of education. I don't know how to thank my dear husband, Jack Green Musselman, he of the keen philosophical mind and generous heart. As a scholar, I am not supposed to let words fail me, but sometimes they do. Thankfully he never does. I dedicate this book to him and to our wondrous new son, Liam.

PART I

Embodied Epistemology

ONE

The Nervous Man of Science

The physiological problem of the formation of our space perception was actually forced upon naturalists by the observation of pathological cases, such as the acquisition of sight in later life through couching, the existence of colour blindness, and a variety of optical delusions which still serve as indispensable test cases for the various theories that have been propounded. Only when something turns out to be palpably wrong do we begin to inquire what constitutes the right side of many things.

—J. T. Merz, *History of European Thought in the Nineteenth Century*

ATOP THE MANLY SHOULDERS of Britain's first industrial age sat some of history's coolest heads of scientific genius: James Watt, Humphry Davy, and William Thomson (Lord Kelvin), just to name a few. While new generations of scientifically and industrially minded men tethered Britain's profits, governance, and other public domains, women disappeared into factories, dark streets, yellow wallpaper, and madhouses.

At least, this is a typical vision of the gendered and scientific quality of Britain's Industrial Revolution. But look a little closer, and several of the narrative's threads show signs of fraying. For example, we have begun to question the extent to which science drove the industrial engine.[1] This book questions a different aspect of the standard story. If masculinity and science epitomized vigor and rationality in industrial Britain, then why and how did so many men of science so frequently endure nervous illness? Among the middling and genteel classes that elite natural philosophers occupied, nervous illness ran rampant, and neither men—nor men of science in particular—were spared from what Elaine Showalter famously called the female malady.[2]

Examples of nervous illness among prominent men and women of science in industrializing Britain are legion. Cambridge dons Richard Watson and Isaac Milner were both hypochondriacs. Watson's teaching duties, he believed, compromised his health, and he stopped giving lectures in 1790.[3] Much to the chagrin of a nineteenth-century hagiographer, James Watt's delicate health seemed to have hindered his abilities in mathematics and the technological arts.[4] German-born astronomer William Herschel's declining health heralded a nervous breakdown and required him to abandon observing for theoretical work in 1802.[5] Chemist Humphry Davy so zealously pursued his famous experiments on the effects of breathing nitrous oxide that he was forced to recuperate in Cornwall, "where new associations of ideas and feelings, common exercise, a pure atmosphere, luxurious diet, and moderate indulgence in wine, in a month restored me to health and vigour."[6] Under less chemically induced circumstances, Mary Somerville also found herself cripplingly fatigued by her translation and explanation of Laplace's *Mécanique Céleste,* and she recuperated in Paris.[7] Charles Babbage became so absorbed in his initial plans for the difference engine that he made himself ill, and his doctor recommended a respite from the project. Babbage's enthusiasm for the project hardly flagged, however, and neither did his periodic nervous complaints.[8] A young James Clerk Maxwell attended school irregularly because of his "delicate health." His attempts to keep up his energy by jogging through the Trinity College corridors in the morning's wee hours annoyed many of his classmates.[9] From his midthirties onward, Herbert Spencer endured a neurotic condition that often prevented him from sleeping, working, or socializing. Ironically, his "nervous system finally gave way" while he was writing the chapter on reason for his *Principles of Psychology,* and prevented him for some time from completing a chapter on the will.[10]

The biographies, correspondence, and even some published scientific papers from the period 1780–1860 are filled with such examples of ill health and physical abnormality among British natural philosophers. These men (and some women) found some solace in writing to each other of their ailments. They also joined their contemporaries in swapping homespun remedies and palliatives, a practice that must have been aided by the chemical experimentation that engaged so many of them.[11] We have mistakenly overlooked the importance of these scientific conversations about nervous disorder, perhaps because they tend to occur in the margins—literally, at the beginning or end of letters, sandwiching the seemingly more important scientific meat—and tend to adopt a mundane tone. We should not be so easily fooled. Men of science experienced nervous abnormalities as significant events not just in their own lives but also in the very fate of science itself.

Our interest in scientists' nervous conditions thus extends well beyond biographical intrigue. These nervous conditions can also tell us a much more significant tale about how science secured its place in modern, industrial society. In this book, I argue that early industrial British natural philosophers thought of the well-maintained nervous system as a model of the ideal scientific and social organization that they hoped to institute nationally and even internationally.[12] The preventatives and palliatives that kept the nervous system in order could have the same salutary effect on the sciences and the rest of society. When working properly, the nervous system literally embodied good scientific method. Simple sensations entered the body at its extremes; those sensations became gradually refined into facts, generalizations, and laws as information traveled through the nerves, into the brain, and finally entered the mind. Ideally speaking, that division of labor was supposed to obtain on a macroscopic scale as well. For example, the fact collectors who populated the city streets and countryside acted as the nerve endings who passed unfiltered information inward to the more mentally sophisticated philosophers at the metropolitan centers. In the physiological and philosophical works of these elite natural philosophers, both nervous physiology and scientific and other social organizations necessarily involved hierarchical systems of management. The very personal and sometimes agonizing experience of illness ironically imbued natural philosophers with the confidence to medicate not only themselves, but also a variety of social ills, with the healing powers of science.

To understand why this is significant, we must understand that Britain in the late eighteenth and nineteenth centuries was undergoing a serious investigation of its own status as a nation and the sciences' place in it. The proliferation of industrial economies, rapid growth of towns, rising importance of provincial areas, and increasing pressure to recognize nonconformist religions all fertilized the political landscape. Many of the nation's leading thinkers (including its natural philosophers) believed that the road forward should be paved with a more extensive national government. The sciences offered a model of rationality and information exchange that would assist with the systematic relief of poverty, religious factionalism, economic inefficiency, and other problems. In the first industrial age, men and women of science dedicated themselves to improving methods for collecting the idiosyncratic experiences of individuals (including individual observers, but also individual towns, counties, religions) and converting them systematically into universally reliable data and theories.[13]

If this system sounds rather mechanical, this is no accident, since most functions of a nervous system, an industrial economy, or a scientific investigation were considered virtually automatic. But before we simplistically contrast soulless, mechanistic philosophers to their Romantic and vitalistic counterparts on the Continent and in poetic circles, we cannot forget how important the mind, free will, and the Creator were to all but the very most radical philosophers in Britain. If the world obeyed natural law, the vast majority of Britain's natural philosophers believed, it did so because of divine design. Similarly, if the body performed an amazing number of functions automatically, this did not preclude the existence of consciousness; if free trade ensured the best economy, the occasional protectionist legislation might help correct its course; if workers or even machines could keep a factory running nearly automatically, one never left them entirely unsupervised. Each of these caveats signaled a strong desire among British elites to preserve human free will and individual moral character.

These analogous wishes to systematize *and* to preserve some independence from automatism required a healthy appetite for paradox and delicate balance. For example, nervous disorders simultaneously threatened to compromise one's scientific authority and masculine integrity, but also promised to verify that authority if one could exert mental control over the disease. To admit to a nervous weakness in the first place, the natural philosopher had to be able to rely upon free will and rationality to keep that weakness in check. The very narrative of restoration foregrounded a weakness in order to display one's strength in subduing it. Anne Hunsaker Hawkins makes this point in her discussion of "pathographies," autobiographical memoirs written by modern-day physicians who suffer an illness: "Pathographies concern the attempts of individuals to orient themselves in the world of sickness . . . to achieve a new balance between self and reality, to arrive at an objective relationship both to experience and to the experiencing self. The task of the author of a pathography is not only to describe this disordering process, but also to restore to reality its lost coherence and to discover, or create, a meaning that can bind it together again."[14]

The pathography, or what Anita Guerrini has called the "spiritual autobiography," is a time-honored genre modeled on Augustine's *Confessions*. The authors of these narratives seek to demonstrate their development from profligacy, through conversion, to salvation. The early modern physician George Cheyne, for instance, experienced chronic vertigo and hypochondria after the emotional distress of his repeated attempts to achieve acceptance among the Newtonians. His illness inspired his widely popular work, *The English Malady*

(1733). In that bildungsroman, he argued that in order to heal the body, one had also to heal the soul, and he sealed this point by recounting his own resurrection.[15] A recognizable genre of pathography or spiritual autobiography only increased in popularity into the nineteenth century.

Peter Melville Logan refers to these texts as "nervous narratives" because a new attention to the nervous system at the turn of the century had changed the nature of disease, particularly among the middle and upper classes. A nervous narrative, he argues, paradoxically "promotes, in its formal structure, the same disorder it cautions against by transforming the narrator's debility into narrative premise. . . . Thus these narratives have to negotiate two contradictory problems, one in which hysteria implicitly undermines the authority to speak, the other in which it becomes the basic condition of speech."[16]

Drawing together these insights, I argue that during the early industrial period in Britain (about 1780–1860) natural philosophers developed special narrative tools for finding meaning in their illness. Like their kindred authors of nervous narratives, natural philosophers turned the adversity of physical weakness into the virtue of the sciences' mental and organizational strength. In turn, that personal apotheosis might serve as a model for the nation's transcendence over its own weaknesses. The mid-Victorian confidence in the sciences, masculinity, and the nation that we have come to take for granted as bedrocks actually had to be rolled, in Sisyphean fashion, up the hill over and over again in the eighteenth and nineteenth centuries. The sciences secured their place in modern society by constantly asserting the power of management over adversity, not by achieving perfect enlightenment. The marginalia describing natural philosophers' illness and recovery, then, turn out to be the frayed edges of a deeply uncertain fabric of scientific, gender, and nationalist politics. The remainder of this book will tug at some of those loose threads in order to unravel what natural philosophers so desperately strove to weave together.

An Overview of the Argument

This book investigates some of the cultural meanings of the nervous system within eighteenth- and nineteenth-century natural philosophy, and the conditions that shaped that meaning. The first part of the book addresses these questions generally. In the remainder of this chapter, I indicate how natural philosophers understood the physiology and philosophy of the nervous system, and why a healthy body mattered so much to scientific practice.

Chapter 2 investigates how subjective experience of both illness and natural phenomena presented problems not only to the sciences but to other areas of society. Natural philosophers sought methods for standardizing experience, and therefore saw themselves as part of the cure for a disjointed society. This cured, reunited body politic was not supposed to be a radically democratic one. When discussing the nervous system, scientific method, and social and scientific organization, natural philosophers nearly always assumed a hierarchical order—one in which a management system remained firmly in place.

Each case study in the second part of the book addresses a particular kind of nervous disorder that plagued natural philosophers in the early industrial period. In each chapter, I discuss how natural philosophers used a variety of cultural resources in their attempts to normalize and control their nervous conditions. In every case, a key physical and cultural concern was the management of idiosyncratic experience. These three chapters focus especially on the visual part of the nervous system and the perceptive faculties of the mind. Since natural philosophers relied so keenly on vision in their work, it stands to reason that it received a disproportionate amount of their attention.

Natural philosophers' investigations of color blindness (chapter 3) tended to emphasize the need to control provincialism. John Dalton wrote the first extended scientific paper on the nature of color blindness early in his career. Color-blind himself, and a Quaker who had spent virtually his whole life in northern England, Dalton's horizon of experience seemed triply limited to his contemporaries. How much more provincial could one be than a man who had virtually no experience of London or the Continent, little understanding of the normal experience of color, and who lived under carefully circumscribed religious codes of dress and language? And yet, gentlemen of science of the next generation lauded Dalton as a genius whose atomic theory provided the possibility of a universal standard for chemistry. The story of Dalton's transcendence over his provincial limitations symbolized what could happen everywhere in British science. Color blindness became a platform for extolling the virtues and methods of transforming idiosyncratic experience into universal knowledge.

Chapter 4 considers hemiopsy, a migraine-like problem that plagued a number of prominent natural philosophers. In studying their own nervous difficulties, these men of science envisioned their bodies as efficient industrial machines overseen by rational mental governors. The natural philosopher had to manage an abnormal body just as he would any other technological device in his service. The mind acted as a regulatory governor (the device that kept steam engines from exploding under pressure). The prevalent analogies drawn between engines and bodies indicated many British natural philosophers' hope that the rational mind could manage virtually any mechanical inefficiency. As natural philosophers became increasingly skeptical in the early nineteenth century that the perfectly efficient machine was possible, they proclaimed the continued importance of thinking managers—God, the mind, the factory supervisor, themselves—who intervened to keep the imperfect machines running.

Finally, scientific research on hallucinations and ghosts (chapter 5) allied with a broader reformist attempt to rein in superstitious, gullible, and sectionalist believers under one rationalized, moderate Anglican roof. The rapid growth of non-Anglican religions in the industrial age seemed to indicate a further descent into provincialism that many natural philosophers wished to avoid. The volatile religious and political atmosphere in early-nineteenth-century Britain rekindled a fear among natural philosophers and physicians that the masses clamored for the irrational, both in its materialist-Jacobin and evangelical forms. A renewed wave of literature that appeared during the early industrial period thus encouraged the rationalizing of apparitions. The dominant message that emerged from these works was that using reason to understand visions made better political and intellectual sense than succumbing to the "superstitious" and "enthusiastic" interpretation of visions as communications with the spiritual world—the latter interpretation frequently associated with radical dissenting groups. Rationalist writers did not deny the possibility of miracles, but did sound quite deistic in their claims that only very rarely did a phenomenon defy natural law. Thus emerged yet another strategy for bringing idiosyncratic knowledge under the more rational governance of natural philosophy.

In short, the three chapters in Part II indicate that natural philosophers equated provincialism, economic inefficiency, and a crisis in faith to nervous disorders. Each of these problems suggested that central authority was dissipating. What had become of Truth if the most elite natural philosophers had imperfect vision, if provincial towns had independent scientific communities, if managers did not guide production, if one God no longer reigned over Britain's factionalizing religious culture? If, as Susan Faye Cannon once claimed, a "Truth Complex" dominated early Victorian scientific culture, this resulted only from extraordinary effort in those and previous decades.[17]

The book's conclusion examines a key midcentury victory for the Truth Complex: namely, telegraphy. Almost immediately compared to the nervous system, the telegraphic network that quickly stretched across Britain and the globe in the 1840s–'60s represented the perceived (if not fully real) healthy exchange and coordination of information that natural philosophers had always hoped to achieve in themselves, their science, and their nation. I conclude with some historiographic reflections on the end of natural philosophy.

Early Industrial British Natural Philosophy as a Crucible

If nervous illness was not strictly a female malady, neither was it primarily an English one, Enlightenment physician George Cheyne's sardonic claims to the contrary.[18] At the very least, the vigorous investigations of nervous physiology and psychology on the Continent shaped much of the work in Britain. Where it is appropriate—and it often is—I include in this study Continental and American scientists who influenced these discussions in Britain. Many had extensive correspondences with British natural philosophers and published important work on these subjects. Furthermore, many Continental and American colleagues had nervous problems of their own. For example, Herman Boerhaave, in the midst of his highly influential work in physiology at the turn of the eighteenth century, reportedly "suffered for six weeks from excitement of the brain, bordering on madness, and characterized by that want of sleep, irritability, and indifference to ordinary interests, which so often appear as harbingers of insanity."[19] At the Pulkovo observatory in St. Petersburg, F. A. T. Winnecke suffered a nervous breakdown just after becoming the vice director in 1864. He chalked up his collapse to the long hours he had kept since he was a student. He resigned from Pulkovo, and rested in Germany for seven years before becoming director of the Strasbourg Observatory.[20] Swiss naturalist Charles Bonnet's eyesight deteriorated so much over his career that he had to stop using the microscope that had made his fame.[21] Joseph Henry and François Arago had hemiopsy (see chapter 4).[22] Jan Purkyne and Gustav Fechner temporarily lost their sight while experimenting on afterimages.[23] John Sylvain Bailly was severely nearsighted.[24] Hermann von Helmholtz, as a migraine sufferer, had personal reasons for immersing himself in the study of afterimages and the physiology of vision more generally.[25]

This is a book about Britain, however, for one simple reason: from 1780–1860, the country experienced unique economic, political, and cultural changes that crucially shaped the sciences' place in society. These broad-based transformations supplied the context and discourse that natural philosophers used to make sense of nervous disorders. British natural philosophers shared a "cultural epistemology," or a core set of assumptions and strategies. One of the major projects toward which natural philosophers turned their shared cultural epistemology was the building of a systematic national identity for Britain.[26] Britain's early industrialization, unique history with Protestantism, and dedication to reform all helped to forge a unique environment for the country's natural philosophers. Post-Revolutionary France harbored a more republican vision of the scientific community as laborers.[27] And while a similar literature on nervous temperament developed in the United States in this period, the different political, economic, and religious context gave this literature a different spin. For instance, the prominent New England psychologist Amariah Brigham averred that the level of democratic freedom in the United States was to blame for the nation's high rate of insanity, which was at least double that of any European nation.[28]

The reader might also wonder why I have chosen to focus on natural philosophers, that is, the array of people who dedicated their time to studying the divinely created universe of "natural bodies, . . . their powers, natures, operations and interactions."[29] After all, many physicians, field naturalists, and technicians also labored under nervous disorders. I limited my subject matter in this way partly because other scholars have written, or are writing, about these other groups. Illness among physicians makes for an especially interesting study, and many other scholars have shed light on health and self-study among doctors in early-modern and modern medicine.[30] A segment of the Darwin industry has focused on the nature of his illness, while Anne Secord has turned our attention toward the embodied nature of epistemologies and methods among a broader swath of naturalists.[31] To the extent that they shared intellectual and social relationships with natural philosophers, I have included in this book a number of instrument makers, engineers, naturalists, and physicians.

However, not only British men and women of science, but more precisely British natural philosophers shared their own cultural epistemology, one that would matter greatly to an industrializing society. For all of their differences on matters of theory, method, and so on, one can detect a loose consensus among British natural philosophers in this period. Certain things mattered: precision, mathematics and instruments as guarantors of that precision, experimental techniques, the structure and dynamics of matter, mind-body dualism as a foundational concept underlying inductivism, divine lawfulness in the universe, and so on. Each of these points of loose consensus played prominently in British natural philosophers' attempts to make sense of nervous disorders.

This is not to say that a highly coherent group existed in Britain that recognized itself as "The Natural Philosophers." For one thing, British natural philosophers disagreed with each other on many important issues. For example, David Brewster, John Herschel, and William Whewell, three natural philosophers who figure prominently in this study, held very different views about government involvement and division of labor in the sciences.[32] In addition, the meaning of natural philosophy changed over time. Enlightenment natural philosophers tended to conceive of nature as a mechanical system of balanced forces where errors would eventually disappear in the return of a pendulum swing. The early nineteenth century saw a shift to a more dynamic, steam-engine-like system that could lose energy and whose anomalies had to be managed rather than erased in order to achieve optimum efficiency. Finally, the mid-to-late Victorian period saw another shift, in which men of science became more materialist, secular, and specialized in their approach to nature.[33]

Despite differences of opinion and changes over time, a relatively cohesive natural philosophy persisted in Britain well into the nineteenth century as a loose term describing those concerned with matter, forces, and the properties of motion and change in nature.[34] By looking broadly at what we would now call astronomers, chemists, physicists, physiologists, philosophers of science, and engineers, I hope to overcome some of the discipline-splitting that threatens to artificially divide the field of history. In many instances, we might fruitfully focus on, say, the history of biology or the history of chemistry, but the farther we peer into the sciences' past, the less sense these modern categories make. Science in the professional, highly specialized sense did not emerge until after the mid-nineteenth century. I try to avoid anachronism by focusing on an issue (in this case, how to handle idiosyncrasy in the nervous system, the sciences, and society at large) rather than a discipline (for instance, physiology). I further discuss the significance of my interdisciplinary approach in the concluding chapter.

Why focus on the early industrial age? Clearly this period saw dramatic changes in many areas of natural philosophy, including electricity, magnetism, matter theory, astronomy, and physical optics—not to mention, as I already have, profound changes in the wider society. But what about nervous physiology? Writing in the early nineteenth century, Joseph Priestley established that the physiology of vision had inspired a great deal of research in the previous hundred years, but there are indications that this interest languished in Priestley's own time.[35] The dramatic advances in experimental physiology made on the Continent only made a serious mark in Britain in the last third of the nineteenth century.[36] Before midcentury, physiological optics—to the extent that it flourished at all—ventured off in numerous, disparate directions. The sciences of nervous physiology before the 1870s in Britain at best enjoyed a "pre-paradigmatic" state.[37]

Early industrial British physiology indeed may not have enjoyed a unifying paradigm in a robust sense. We can again, however, identify a core cultural epistemology that gave coherence to investigations of the nervous system. In particular, as I have already mentioned, this research was shaped by natural philosophers' desire to organize idiosyncratic experiences under rational theories, practices, and philosophies of knowledge. National governance and reform, political economy, and rational religion provided some of the idioms through which natural philosophers understood and managed nervous physiology at both the personal and scientific level.

Just as we would be mistaken, though, to homogenize British natural philosophers' views, we also cannot ignore the profound changes that occurred within this community over time. The early industrial period grappled with the legacy of Enlightenment and the insights of Romanticism, and forged mid-Victorian confidence. I will trace these historical changes in more detail in later chapters, but we can summarize the trajectory this way: first, the Enlightenment bequeathed to the late eighteenth century both the promise of systematic, unified knowledge grounded in the System of Nature and the thorny problem of how to incorporate local peculiarities into one grand narrative. Early-nineteenth-century philosophies immersed themselves in this tension between local and universal and attempted to resolve it.[38] I argue that we can see this tension in natural philosophers' attempts to deal with their own nervously disordered bodies. Empiricism had become foundational to British Enlightenment thought, but empiricism required a fully functioning nervous system, a luxury many men of science did not enjoy.

Early industrial British natural philosophers envisioned a way out of this problem by making a virtue out of necessity: nervous disorder would provide the opportunity to develop stronger discipline and will. Knowledge would come not through spontaneous revelation or enlightenment, but through carefully structured labor. The challenge posed by the idiosyncratic and the local was met by natural philosophers with the solution of management. Rather than eradicating imperfection, natural philosophy instead sought to extract as much work from imperfect systems as possible. Conceding bodily and mechanical imperfections and yet achieving work regardless made victory all the sweeter. In fact, the more that natural philosophers studied how to achieve the most efficiency from bodies and machines over the first half of the nineteenth century, the more confident they became that the idiosyncratic and the local really posed no serious problem to the structuring power of the sciences. Mid-Victorian scientific confidence in management remained firmly grounded in the earlier dilemma—even in this heyday of optimism about objectivity, the idiosyncrasies of the self never disappeared from the picture—but relegated the unruly psyche to the background and to the interior.

Who exactly were these early industrial British natural philosophers who populate this study, then? I have identified a loosely knit and multigenerational set of practitioners who shared a need to organize idiosyncratic experiences into systematic knowledge *and* an interest in the nervous system, often because their own was out of whack. The actual social ties that bound these men, and a few women, together were complex. Most of my subjects knew each other through universities (especially Cambridge and Edinburgh), through scientific societies (especially the Royal Society, British Association for the Advancement of Science, and provincial groups), and in some cases through government bodies (such as the Greenwich Observatory and the Board of Longitude). Most of my protagonists identified with a moderate whig politics for most of their careers. Most came from financially comfortable families, some of them titled. All believed in the Divine, and if they were not Anglican, they at least played their nonconformism in a minor key (with the notable exceptions of Joseph Priestley and David Brewster).[39] When their differences threatened to rend their social fabric, they fell back on a common commitment to instituting the sciences as society's most reliable adhesive. In the nervous system and its disorders they found a useful arena for debating how precisely the sciences would help structure and cohere that society.

The Philosophy and Physiology of the Nervous System

Natural philosophy in the eighteenth and nineteenth centuries owed a great deal to contemporary theories and experiments on the body and mind. In particular, scientific epistemology and method frequently looked to nervous physiology to learn what was possible for human investigators.[40] In the industrial age, anatomists, physiologists, and philosophers devoted substantial attention to understanding the nervous system and how it might structure thought.

The ideas of the philosophers and physicians of the Scottish Common Sense school set the agenda for subsequent British mental and moral philosophy. The Common Sense philosophers, for instance, made an important distinction between sensation and perception. Sensation was the virtually unmediated mapping of the outside world onto the retina, while perception was the mind's active judgment of what the sensation was and what it meant.[41] This distinction helped differentiate the material and passive aspects of nervous physiology from the immaterial and dynamic nature of mental activity. A healthful life and science therefore required not only a well-kept body, but also the cultivation of mental and moral faculties such as reason, judgment, heightened attention, sympathy, and sensibility.

In his 1749 *Observations on Man,* English physician David Hartley had made the explicit, associationist connections between nervous physiology and the mental-moral philosophy that would so dominate British thought. He argued that ideas arose and became connected through vibrations in the nerves and brain.[42] Following on Hartley's example, two of the most important early figures in the Scottish Enlightenment, William Cullen and Robert Whytt, had given their colleague's mental philosophy a more empirical basis in nervous physiology. Their approach emphasized the active nature of the mind—and therefore, the active nature of perception and other mental processes. Furthermore, they elevated the nervous system to the place of chief importance in physiology. Cullen informed his students at the University of Edinburgh that the nervous system, "as the organ of sense and motion, is connected with so many functions of the animal oeconomy, that the study of it must be of the utmost importance, and a fundamental part of the study of the whole oeconomy."[43] Popular texts through the first half of the nineteenth century reinforced the idea that the nerves acted not only as a route for sensations from the outside world, but also as the conduit for the mind's direction of the body. For physician John Elliot, for example, the human body was "a machine composed of bones and muscles, with their proper appendages, for the purpose of motion at the instance of its intelligent principle; from this principle nerves, or instruments of sensation, are likewise detached to the various parts of the body, for such information as may be necessary for determining it to those motions of the body which may be most conducive to the happiness of the former, and preservation of both."[44] Members of a later generation such as Charles Bell continued to challenge the passive model of the nerves maintained by Albrecht von Haller and other materialists. Bell insisted that the mind was not merely acted upon, but active during sensation and other nervous activities.[45]

This idea made sense in light of the connections formed in the early nineteenth century between nervous impulses and other imponderable (invisible) forces. If nervous impulses acted like electricity, for example, it stood to reason that the nerves existed in a continually active state. Luigi Galvani's connection between electrical and nervous impulses and Johannes Müller's law of specific energies both emphasized the the body's similarity to a machine powered by imponderable forces. Just as varying stimuli (e.g., electricity, mechanical pressure) could garner the same response when applied to a nerve, so also machine technology was demonstrating the correlation of different kinds of forces.[46] By the mid-nineteenth century, London physician Henry Holland considered it a commonplace to liken nervous power to light, electricity, magnetism, heat, and chemical force. Important implications of this analogy for Holland were the continuity of material and mental phenomena, and the possibility that the rational will might still control the "more automatic machinery which surrounds it."[47] By the time Holland was flourishing, this image of the body as a machine governed by a mind had become quite popular. Not only epistemological idealists such as William Whewell, but also many of his empiricist critics argued that the mind actively shaped knowledge of the world.[48]

Interest in the connections between mental philosophy and nervous physiology was widespread within the natural philosophical and medical communities. A quick skim of any scientific periodical from this period demonstrates a keen fascination with bodily abnormalities and their effects. Goethe's *Zur Farbenlehre* (1810) and Brewster's *Letters on Natural Magic* (1832) are only the two best-known treatments of these phenomena. Speculations and experiments on all manner of illusions and aches, phantasmagoria and pangs, appeared in the specialized and popular scientific literature. The *Philosophical Magazine* published an especially large number of notices about various nervous effects—understandably, given that that prolific student of optics David Brewster edited the journal from 1832 to 1868.[49] But more formal society transactions also appeased their readers' interest in experiences of nervous malfunction and its effects on the understanding.[50] In their correspondence and private notes as well, natural philosophers noted their experiences with nervous disorder.[51]

In the last quarter of the nineteenth century, Hubert Airy (son of Greenwich Observatory director George Airy) looked back over this vast literature and proclaimed natural philosophers especially qualified for the study of nervous disorder. Because of their particular education and experience, he argued, natural philosophers had unique claims to authority on the subject that surpassed even the claims of physicians:

The votaries of Natural Philosophy are especially qualified by their habits of accurate observation to contemplate attentively any strange apparition, without or
within, and, I had almost said, are especially exposed to the risk of impairment
(temporary or permanent) of the eyesight, by the severity of the eye-work and
brain-work they undergo, and therefore possess especial advantages for the
study of visual derangements; whereas the physician, unless personally subject
to the malady, must depend, for his acquaintance with its phenomena, on the
imperfect or exaggerated accounts of patients untrained to observe closely or
record faithfully.[52]

According to Airy, anyone who presumed that the physiology of the nervous
system was the exclusive territory of physicians, thought wrong. Nervous
physiology—and particularly vision—played too important a part in the methodology of natural philosophy. The natural philosopher not only valued the
nervous system as a tool; he also often worked it to the point of impairment.
Finally, he brought the necessary training to investigate accurately any physiological problems he might himself have. His familiarity with epistemology, the
science of imponderable forces, precision instrumentation, and optics made
him a unique authority on nervous issues. Ironically, a natural philosopher's
bodily vulnerability afforded him the heroic opportunity to achieve mastery.
Masculine scientific "habits" constituted the endless work of keeping the house
in order.

VISION IN NATURAL PHILOSOPHY

For natural philosophers especially, one of the creakiest, but most vital parts of
the house was the visual system.[53] Accordingly, natural philosophers had a
particular interest in those parts of the nervous system and mind involved
with vision. It is therefore worthwhile to consider the particular epistemological problems and solutions raised by that sense and its disorders. Like other
kinds of perception, vision seemed to engender an unavoidably subjective experience, though the fact that the vast majority of people gave very similar descriptions of everyday phenomena generally masked this subjectivity. Problems
arose when one tried to explain the experiences of people who saw phenomena
that others could not see (e.g., the spots one sees after looking directly at the
sun, or worse, hallucinations), or who gave significantly different descriptions of

what should have been the same phenomenon (e.g., observers' differences about what time Venus begins and ends its transit across the sun).

Despite a trend toward the use of self-registering instruments, the practice of natural philosophy still depended greatly on human perception. And for all its faults, vision held a privileged place among the senses. Sight seemed to provide the natural philosopher with the most, and the most accurate, information. Chemist and physician Samuel Brown vividly captured the importance of vision to the natural sciences:

> It may be said that it is always the first effort of the exact sciences to transform the dimmer perceptions of the more deceivable organs into those of sight, the most discursive and accurate of the senses. The mineralogist does not satisfy himself with the intimations of what has been called the muscular sense [touch], or that sense of resistance which is related to the perception of weight, concerning the specific gravity of the stone. He weighs it first in the air, then in the water; notes the difference between the two weights; and thence computes its specific heaviness. The chemist does not trust his fingers, or even his lip, for the temperature of his agents and reagents; but invents the thermometer, and reads of his measurements with the eye. It is the same in the sciences of magnetism proper, electricity, and galvanism. Even in the investigation of sound (which is measurable with such exquisite nicety by the ear, as to render the art of music not only possible, but the very anti-type of mathematical proportion,) the natural philosopher converts its vibration into visible things before he will philosophize upon them.[54]

Just after extolling the virtues of vision, Brown warned that it should still be checked by more trustworthy instruments such as micrometers and photoscopes. In fact, a careful examination of the quotation above indicates that Brown privileged sight more because it was the main point of contact between instruments and the mind, and less because of any inherent superiority to that sense. Indeed, natural philosophers did not feel universally comfortable with their dependence on their visual capacity. French *physicien* Eugène Péclet instructed readers of his textbook that "the imperfection of our organs and of our instruments does not allow us to make absolutely exact observations; they will never rigorously satisfy the laws that govern them; one must require only that the differences be smaller than the probable errors in the instruments. Further, the series of observations must be very extensive; for one would rather risk obtaining, not a general law, but a law that would apply only within the period observed."[55]

The careful observer attended not only to the facts gathered by sensation, but also to deduction, which could be a powerful corrective for perceptual error.[56] Once a general law was established it could be used to isolate and eliminate such errors. But notice the persistent uncertainty in these passages. Eyesight did not promise perfect knowledge, only *better* knowledge than the other senses offered. Likewise, empirical methods could lead to powerful, general scientific laws, but not absolute certainty. Mastery required constant labor and discipline. In short, natural philosophy, one of the cornerstones of modern science, rooted its power in its greatest source of vulnerability. It promised progress through visionary work, not a static truth.

Warnings about visual imperfection became more urgent as the accuracy of instruments and theories improved. Perceptual errors appeared more and more gross by comparison to the fine registers of precision instruments. Even healthy human perception could not always be trusted to produce accurate accounts of natural phenomena. Any lens had its imperfections, and as a special case of the lens, the eye came with its share of problems. For example, the constant, involuntary adjustment of the pupil made judging the relative brightness of stars extremely difficult.[57] In some cases, natural philosophers replaced the eye with an instrument such as the camera. In others, they compared the sightings of several different observers, as with the transits of Venus. In those instances where visual evidence was unavoidable but problematic, natural philosophers sought to discipline vision through various protocols.

Refining these protocols depended partly on improving knowledge about the philosophy and physiology of perception, and partly on the sheer quality of the scientific investigator's higher mental faculties. In other words, mechanization alone could not correct for the flaws in the body's perceptual apparatus. The natural philosopher had also to cultivate conscious mental muscles such as the will and attention. These would help him accurately to interpret raw data from his senses and instruments. Both in the body and in the system of scientific methodology, nervous physiology and empiricism thus had far more complexity than simple mechanical processes.

Attention, the Will, and Power

British mental and moral philosophy in this period tended to highlight the mind's various faculties, or the types of operations that the mind could perform. Intense debates ensued as to whether these operations were metaphysical entities or simply heuristic conveniences. Phrenologists and many other anatomists assigned these faculties to specific areas of the brain, but even those who viewed such reification as speculative agreed that the mind and brain had a number of distinct ways of processing information.[58] In the empiricist-associationist view that dominated British thought, some faculties enabled action (e.g., the will, appetites, and the moral faculty), while others enabled understanding (e.g., memory and judgment). Of the faculties of understanding, some gave trustworthy accounts of the external world (e.g., perception, abstraction, and reason), while others transformed that world into something generally unusable to natural philosophy (the imagination).

In an account of nervous disorders and scientific epistemology, the faculty of perception of course played a very important part. But I also want to emphasize the vast importance of the will and attention in empiricist accounts of the mind and body. These "active" faculties trumped any radical moves that British philosophy might have made toward total mechanization or skepticism.[59] For example, David Hume famously attacked the robust notion of causation that buttressed much moral and natural philosophy. Seeking to moderate this skepticism, empiricist philosophers in the next few generations appealed to a commonsense experience. When human beings will an arm to move, said Thomas Brown in an oft-repeated argument, they can see it move, and simultaneously feel the exertion of the muscles lifting the arm. This visceral sense of a cause powering an effect allowed one to believe that at least human action had a clear cause. One willed and felt the force of that will simultaneously, ergo causation.[60]

The will's power mattered in early industrial British society partly because it had become so important in differentiating masculinity and femininity. During the eighteenth century, women had become increasingly associated with sensibility, or nervous rawness of feeling. Men were also supposed to cultivate sensibility, but most believed that men could control its wilder impulses through the exercise of higher mental faculties. In the industrial age, though, balancing an active life and work ethic against the virtues of contemplation and sensibility proved no simple matter. Even the most seemingly confident believer in the power of heroes, Thomas Carlyle, in reality had tremendous difficulty carving out a suitably active, masculine life for himself as a writer.[61]

In early-nineteenth-century natural philosophy as in letters, the assertion of masculine will became a crucial ideal, an ideal honored more in the breach than in the observance. The will would, natural philosophers hoped, transform the feminine, hypersensible experience of nature's beauty into a vigorous, exhilirating wrestling match. The much-respected mid-Victorian natural philosopher James Clerk Maxwell vividly painted this image for his colleagues at the British Association for the Advancement of Science:

> There are [those] who feel more enjoyment in following geometrical forms, which they draw upon paper, or build up in the empty space before them.
>
> Others, again, are not content unless they can project their whole physical energies into the scene which they conjure up. They learn at what rate the planets rush through space, and they experience a delightful feeling of exhilaration. They calculate the forces with which heavenly bodies pull at one another, and they feel their own muscles straining with the effort.
>
> To such men momentum, mass, and energy are not mere abstract expressions of the result of scientific inquiry. They are words of power, which stir the souls like the memories of childhood.[62]

Maxwell sought to retire the image of the natural philosopher as passive camera obscura who projected his geometries into empty space. Into the gap vacated by this weakened, feminized projector, Maxwell thrust the dynamic, physical man of science who rushed into nature and muscled it around instead of simply mirroring it.[63] Over the course of the industrial era, natural philosophers increasingly depicted the Enlightment mirror of nature as inadequately static. The new, heroic vision of natural philosophy required men of action, who professed to *manipulate* nature in a dynamic tug-of-war between sensing and willing, between body and mind.

Maxwell's heroic vision of an active, analytical dynamics would have pleased many of his predecessors, especially William Hamilton, whose lectures on mental philosophy at the University of Edinburgh first presented Maxwell with "the doctrine of a muscular sense [which] gave promise of a rational analysis of the active powers."[64] The will allowed the natural philosopher to do more than simply *observe* and *measure* nature. One threw in one's very body, and *manipulated* and *experienced* nature. As I will argue, particularly in chapters 2 and 4, this highly masculine approach also allowed natural philosophers to manage their own bodily abnormalities, to convert weakness to strength rather than sink into victimhood.

The will proved so important to the practice of science and everyday life, in fact, that doubts about its existence caused several famous nervous breakdowns—including that of the young John Stuart Mill, who suffered his much-discussed crisis in the winter of 1826–27 after immersing himself in the study of

Benthamite utilitarianism. Among his critiques of his father's philosophy was that it was radically deterministic. "I felt as if I was scientifically proved to be the helpless slave of antecedent circumstances; and as if my character and that of all others had been formed for us by agencies beyond our control, and was wholly out of our power." He eventually saw his way through the problem by deciding that free will was compatible with a nonfatalistic view of how circumstances shape moral decisions.[65] Apparently, looking the experience of "slavery" straight in the face could bring one to the brink of disaster but then to an even stronger state of grace.

Related to the all-important will, the faculty of attention also played prominently in moral and natural philosophy. Without attention, the mind remained unable to prioritize or recall ideas. As physiologist William Carpenter put it, "it is solely by the Volitional *direction of the attention* that the will exerts its dominion; so that the acquirement of this power, which is within the reach of every one, should be the primary object of all mental discipline." Those who did not exercise their will by disciplining their attention were little better than automata (or women).[66] Because attention connected the will to the intellectual faculties, it kept the intellect from being strictly mechanical. Sensation, for instance, might consist merely of an automatic impression of light upon the retina, but the mind's attention to certain aspects of that impression made sensation subject to the will. The more powerful one's will, the more developed one's intellectual faculties. The will improved mainly through its exercise in active faculties such as the appetites, desires, and morality. Therefore, power, morality, and the intellect all connected through the nexus of attention.

Natural philosophers repeatedly credited attention as one of their sharpest disciplinary tools. Among the many innovations that George Airy introduced to the Royal Greenwich Observatory, for instance, was an alarm clock set to sound whenever certain stars crossed the meridian. This alerted the observer on duty to check his instruments.[67] Charles Babbage, reflecting on his success in life, also extolled the importance of well-directed attention. One of his "most important guiding principles," he said, was that "every moment of my waking hours has always been occupied by some train of enquiry." Sometimes this meant working in the wee hours of the morning, one of the only times he found respite from "the nuisances of the London streets," most particularly their organ grinders.[68] Echoing Hubert Airy's point (cited earlier) that natural philosophers had the best firsthand knowledge about flawed vision, Hermann von Helmholtz also sounded this common theme when he argued that focused attention was rare but definitive of the best natural science:

Who can easily discover that there is an absolutely blind point, the so-called *punctum caecum*, within the retina of every healthy eye? How many people know that the only objects they see single are those at which they are looking, and that all other objects, behind or before these, appear double? I could adduce a long list of similar examples, which have not been brought to light till the actions of the senses were scientifically investigated, and which remain obstinately concealed, till attention has been drawn to them by appropriate means—often an extremely difficult task to accomplish.[69]

If brought under rigorous investigation, even the natural philosopher's own physical abnormalities might become legitimate subjects of scientific knowledge. Active mental faculties could turn even perceptual errors into object lessons on how to improve scientific practice. Given the realities that instruments never achieved perfect accuracy, eyesight could play tricks, and bodies fell ill, natural philosophers relied on their minds as a beacon through the fog.

The Body in Scientific Practice

The fact that the body intruded so unpredictably and demandingly into the practice of science is one of the primary concerns of this book. Since investigation of the body's cultural importance has arrived rather late to science studies, I want to address briefly how historians have talked about the body in recent years, and discuss more thoroughly the body's place in the sciences.[70]

The historical literature on the body as a cultural force began to appear in the 1970s–'80s, following particularly on the work of Michel Foucault and Norbert Elias, who sketched the body as a locus of power relations within the construction of modernity,[71] and feminists such as Evelyn Fox Keller, Genevieve Lloyd, and Susan Bordo, who documented the historical division of labor between embodied women and intellectual men.[72] Both sets of literature emphasized that the body, long taken for granted as a hard-wired instrument, had absorbed as many cultural values as the more familiarly flexible icons of gender, race, and class.

One part of this literature has presented us with a mundane and yet simultaneously profound fact: illness, and the management of illness, were regular features of life before our own time.[73] Bruce Haley vividly illustrated this point in his study of the Victorian culture of health. He argued that the Victorians imbued the healthy body with enormous cultural meaning. To them, health was "a state of constitutional growth and development in which the bodily systems and mental faculties interoperate harmoniously under the direct motive power of vital energy or the indirect motive power of the moral will, or both. Its signs are, subjectively recognized, a sense of wholeness and unencumbered capability, and, externally recognized, the production of useful, creative labor. All of this is said more simply in *mens sana in corpore sano.*"[74] In other words, the strenuous, unceasing moral and physical task of maintaining personal health also helped one achieve one's social duties: economic productivity, political progress, and piety.

This intimate relationship between health, intellect, productivity, and morality has clear implications for the history of science. Probably because of its deep roots in intellectual history, however, historians of science have been slow to consider the body as much more than a biomedical specimen. In the last two decades, a number of studies have begun to address the issue. The earliest literature considered medicine as a tool that inscribed social, cultural, economic, and religious values onto the body.[75] Another set of studies has examined how bodies have been compared to machines.[76] More recent inquiries have explored how scientific investigators' bodies have shaped scientific practice.[77]

A study of nervous disorders among natural philosophers must involve all of these approaches. In fact, it seems that science studies sits on the verge of a new synthesis in its analysis of the body. Roy Porter expressed the need for this synthesis more than a decade ago: "We need a thick-textured study of the body, unprejudiced by timeless philosophical dualisms or Lovejoyan unit-ideas . . . research which contextualizes the human frame within specific sociocultural frames of reference, sensitive to experience, representations, and meaning."[78] The body has never been just a machine, just an organism, just an instrument, just a device for gender, class, or racial politics. It has been all of these things at once and has undergone constant shifts in meaning. Recognizing this gives us one escape route from the fruitless dichotomy between positivism and constructivism. We do not need to think of the embodiment of science as meaning the reduction of the natural world to just words or just things.[79]

Within the maelstrom of meanings imprinted on the industrial-era body, men and women of science faced a special problem. They had invested in the body as a powerful source of scientific and medical knowledge (both in its capacity as a subject and as a scientific instrument). Yet, all classes of society, including natural philosophers, were subject to physiological idiosyncrasies. If we wish to understand how scientific knowledge has been produced, then, we must pay closer attention to the poor health and body consciousness of its practitioners.

So much of scientific methodology emphasized the desirability of a fully functioning and virtually invisible body. But illness was so common that rendering the body invisible—or at least translucent—involved a great deal of work. The discipline of the natural philosopher's body was a never-ending project. In fact, industrial-era natural philosophers based their claims to accuracy, precision, objectivity, and other scientific virtues on the proper *management* of the body, not the perfection of the body in the first or last place. These men and women aspired to something like the public image of Cambridge Lucasian Professor Stephen Hawking, who is immobilized by Lou Gehrig's disease. As Hélène Mialet has aptly written: "We glorify him because he has transcended the conditions imposed on him by his own body, while the prevailing ideology promotes a scientist without a body or self-awareness."[80]

Even without the complications of near total paralysis, the pursuit of natural knowledge could be punishing work. Numerous medical treatises in the early modern period had emphasized the care required for the upkeep of a scholar's body. Francis Bacon, for instance, outlined two paths to longevity. The first was a "country life" where one's actions were "free and voluntary." Superior even to this lifestyle was that of monks and philosophers who lived under "regulation and commands within themselves; for then the victory and performing of the command giveth a good disposition to the spirits."[81] By contrast, a century later, Cheyne famously dubbed nervous frailty an "English malady" that preyed on contemplatives. He advised such people frequently to shave their heads and faces, wash and scrape their feet, and pare their toenails.[82] Both the lusty poet and the patient philosopher were prone to physical degeneration, acute in the former case, chronic in the latter:

> Your Men of *Imagination* [poets] are generally given to *sensual* Pleasure, because the Objects of *Sense* yield *them* a more delicate *Touch*, and a livelier *Sensation*, than they do *others*. But if they happen to live so long (which is hardly possible), in the *Decline* of Life they pay dearly for the greater bodily Pleasures they enjoyed in youthful Days of their *Vanity*. Those of *rigid, stiff* and *unyielding* Fibres, have *less vivid* Sensations, because it requires a greater Degree of *Force* to overcome a greater *Resistance*. Those excel most in the *Labours* of the *Understanding* [philosophers], or the *Intellectual* Faculties, retain their *Impressions* longest, and pursue them farthest; and are most susceptible of the slow and lasting *Passions*, which secretly consume them as *chronical* Diseases do. And *lastly*, those whose *Organs* of *Sensation* are (if I may speak so) *un-elastick*, or intirely *callous, resty* for want of Exercise, or any way *obstructed*, or naturally *ill-formed*, as they have scarce

any *Passions* at all, or any lively *Sensations*, and are incapable of lasting *Impressions;* so they enjoy the *firmest* Health, and are subject to the fewest *Diseases:* such are *Ideots, Peasants,* and *Mechanicks,* and all those we call *Indolent People.*[83]

Ironically, Cheyne concluded, the "indolent" enjoyed the best physical health. They did not wear down their nervous fibers quickly like the "men of imagination" or slowly like the philosophers.

No one was more convinced of Cheyne's wisdom than one of his patients, David Hume. From the spring of 1730 onward, Hume's work caused him severe emotional despair. Despite Cheyne's pessimism about the health of scholars, Hume set about healing himself by improving his temper, will, reason, and understanding. This might have worked quite well, had he lived an active physical life. Such a life, Cheyne would have advised, would etch his virtuous reflections deeply into his soul. But in his sedentary solitude, those reflections would "serve to little other purpose than to waste the spirits, the force of the mind meeting with no resistance, but wasting itself in the air, like our arm when it misses its aim."[84]

These concerns about the nervous susceptibility of contemplative men intensified in the nineteenth century. Where once nervous illness was associated almost exclusively with the aristocracy, it became a middle-class epidemic in the industrial age. As early as 1768, the physician William Smith proclaimed that "there are very few disorders, which may not in a large sense be called nervous," and by another, later estimate, nervous disorders accounted for two-thirds of all disease in the early nineteenth century.[85] Whether or not this statistic was exaggerated, scholars and professionals could not help but wonder when, not whether, they too would confront nervous illness. James MacKenzie, a founding physician at the Worcester Infirmary, warned his fellow scholars to "endeavour to repair by their temperance, regularity, and care, what is perpetually impaired by their weakness, situation and study."[86] Likewise, John Herschel admonished his friend, the astronomer and stockbroker Francis Baily, to follow his physician's advice while recovering from a concussion. He should stick to light reading, for one could imagine "how hard a task thinking must impose on the nerves and fibres in their delicate offices they have to perform as ministers of the soul."[87]

A key antidote here was *management*, what would become the "therapeutic watchword" of the last decades of the nineteenth century.[88] The well-balanced man of science neither entirely transcended his body nor entirely subordinated himself to it. The work to improve one's health mattered more—and appeared far more realistic—than the actual *achievement* of perfect health. Industrial-age Britons aspired to progress more than perfection.

For those melancholics (such as scholars) who did not physically labor for a living, exercise stood in as a substitute. It set the animal spirits in motion and reinvigorated the body and the sensorium. Conventional wisdom held that physical weakness and nervous complications went hand-in-hand. Samuel Taylor Coleridge, for example, once speculated in casual conversation that color blindness might "proceed from general weakness, which will render the differences [in colors] imperceptible, just as the dusk or twilight makes all colours one."[89] Physician James Jurin noticed that when people aged or neglected to exercise their vision, their eye muscles lost elasticity just as any other unused muscle would. Aging and looking extensively at distant objects led to farsightedness. Reading, microscopy, and other activities which involved looking at near objects produced nearsightedness.[90]

Besides exercise, impeccable moral caliber could also shield the scholar from nervous disorders. As in Charles Bell's division of the nerves into irritable and motor types, the mind also had its receptive and active faculties. From the late eighteenth century, the rationalist approach held that those whose active powers of reason, judgment, and attention held the imagination in check, enjoyed greater intellects and nervous immunity than those who succumbed to exceeding sensitivity, or "sensibility." While training a new generation of philosophers at the University of Edinburgh, Dugald Stewart approvingly quoted his mentor, Thomas Reid, on the subject: "A person of acute sensibility is so much affected with his own strong sensations, produced by the contemplation of any object, or work, calculated to excite them, that he cannot exert any discerning power, which a man of less lively sensations employs in contemplating works of taste."[91]

Those rare individuals who combined reason, regular exercise, good diet, and moral discipline had gathered all the available inoculations against nervous disorder. Jean-Baptiste Biot's 1829 portrait of Newton as subject to psychological instability therefore seemed unfathomable to many British scholars. Far from meaning any disrespect to Newton, Biot figured that a life of continuous and lofty meditations would naturally fatigue any mind. But in a spirited defense of his countryman's sanity, David Brewster argued that the "unbroken equanimity of Newton's mind, the purity of his moral character, his temperate and abstemious life, his ardent and unaffected piety, and the weakness of his imaginative powers, all indicated a mind which was not likely to be overset by any affliction to which it could be exposed." Others, such as William Whewell and Baden Powell, accepted that genius sometimes engendered madness, and simply hoped such affiliations were rare.[92] Either way, the importance of Newton's self-control to natural philosophy's nationalism was clear.

Regardless of their assessments of Newton's character in particular, most British natural philosophers in the industrial era acknowledged that the scholar had to work not to buckle under the weight of his own thoughts. If few wished to question a national hero's sanity, the physical dangers of the scientific life were readily admitted. In fact, thwarting those dangers made the natural philosopher's life all the more heroic. Charles Wheatstone neatly encapsulated this tension between, on the one hand, the desire for the expansion of knowledge, and, on the other hand, the danger to the natural philosopher's health should this desire be overindulged. After examining Czech physiologist Jan Purkyne's ingenious but physically damaging experiments, Wheatstone found it undeniable that "their frequent repetition may be attended with dangerous effects on the eyes. On the other side, it is indispensable that the experiments should be frequently repeated and varied; for at the commencement of the inquiry the observer must be quite unaccustomed to this new field of experiment." Perhaps, Wheatstone demurred, with a hint of post hoc justification, Purkyne's eye problems had really been congenital rather than induced by experimentation.[93] Here, and in other commentaries on the risks of scientific investigation, we can detect an overarching value: the more fundamental the research, the greater the justification for self-sacrifice.

Few natural philosophers yet had the luxury of devoting their full day to the sciences. For many, then, scientific investigation was a relaxing leisure activity that could temper the strain of one's other duties, so long as one did not overindulge. For instance, William Henry lauded his fellow chemist Joseph Priestley for recommending "experimental philosophy as an agreeable relief from employments that excite the feelings or over-strain the attention," and proposed it "to the young, the high-born, and the affluent, as a source of pleasure unalloyed with the anxieties and agitations of public life."[94] The trick was to walk the narrow line of health between work and rest. French *physicien* François Arago characterized John Sylvain Bailly's instruction under Nicolas-Louis de Lacaille as particularly grueling. The modern astronomer, Lacaille told his student, agreed to devote his whole attention to his work, disregarding foul weather or fatigue:

> To complete the observation, he must read off the microscopical divisions of the graduated circle, and for what opticians call *indolent vision* (the only sort the an-

cients ever required) must substitute *strained vision*, which in a few years brings
on blindness.

When he has scarcely escaped from this physical and moral torture, and the
astronomer wishes to know what degree of utility is deducible from his labours, he
is obliged to plunge into numerical calculations of a repelling length and intricacy.

The English translators of Arago's éloges took umbrage with his complaints
about the work required in astronomy, and expressed admiration for Lacaille's
"very great practical perseverance."[95] This disagreement, however, should not
be read as indicating a national difference of opinion over the strenuousness of
astronomy. Arago's emphasis on the physical trials of his éloge subjects was not
unusual to biographical writing either in France or Britain. The significance of
the English rebuttal rather speaks to the confusion among natural philosophers
about what levels of diligence would best promote *both* the growth of knowledge *and* personal and national health. This point was vividly and tragically illustrated by the suicide of William Henry, the same Manchester chemist
whom we heard earlier thanking Priestley for offering experimental philosophy
as a relief from public life. In the 1830s Henry became more involved with or-
ganizing meetings of the British Association for the Advancement of Science,
a highly public and national exercise of experimental philosophy. Several historians have argued that the extreme stress Henry endured in Bristol was the
proximate cause for his suicide.[96]

While the extremity of Henry's suffering was the exception rather than
the rule, we have seen how common ill health and particularly nervous disorders were among British natural philosophers. Because the nervous system also
occupied the center of the inductive process in the sciences, the natural philosopher strove after nervous health not only for physical comfort, but also for the
sake of epistemological authority. The active and unceasing cultivation of a rational mind and a healthy body thus constituted a normal part of the scientific
process. Natural philosophers considered their arduous but successful management of the body as one indication of their ability to manage other areas of
British life. As Hubert Airy would argue by 1870, the very physical frailties that
threatened to bring natural philosophy's authority to its knees, actually gave
the man of science the opportunity to exercise his peculiarly cultivated, disciplined habits of mind. Or so they hoped. As we will continue to see in subsequent chapters, industrial-age natural philosophers' proclamations of self-
mastery and authority to govern other national affairs had to be established,
not assumed.

TWO

The Social Hierarchy of Subjectivity

E'en with respect to human things and forms,
We estimate and know them but in solitude.
The eye of the worldly man is insect-like,
Fit only for the near and single objects;
The true philosopher in distance sees them,
And scans their forms, their bearings, and relations.
To view a lovely landscape as a whole,
We do not fix upon one cave or rock,
Or woody hill, out of the mighty range
Of the wide scenery, — we rather mount
A lofty knoll to mark the varied whole.

—Humphry Davy, quoted in John Davy,
Memoirs of the Life of Sir Humphry Davy

NERVOUS DISORDERS DISRUPTED the day-to-day life of natural philosophy and threatened the tenuous claim that natural philosophy had to the Enlightenment crown of reason. Generalizing from one's own experience of nature meant transforming that experience into something that was not utterly idiosyncratic, that could be understood by others, that could become Knowledge. In fact, this project of converting individual experience into communicable knowledge confronted not only those with nervous disorders, but every student of the sciences. The epistemological difficulties faced by natural philosophers with nervous disorders differed from the difficulties of everyday science more in degree than in kind.

If early Enlightenment observers assumed that individual subjectivity would eventually wither away under the light of science, post-Romantic generations lacked that certainty. In fact, the possibility presented itself that no objective world existed beyond individuals' subjective experience. This possibility was anathema to a British scientific community that overwhelmingly endorsed a more integrated science and stronger nation. Where Enlightenment sciences treasured experience, autonomy, progress, and emancipation, the generations who followed the French Revolution felt the need to re-center and reestablish calm. *Expertise* would become the new watchword in a scientific culture that "accepted the necessity of social inequality and inescapable limits on material progress."[1] From the early nineteenth century, natural philosophers built a system of expertise by developing strategies to manage the peculiarities of individual observers. Instruments could offer more accurate and precise representations of phenomena. For those tasks that still required human intervention, training could reduce subjectivity. The mathematics of error analysis quantified the probability that an individual observation was correct.[2] As discussed in the previous chapter, autobiographical nervous narratives could also discipline one's subjective experience.

Men and women of science were not the only ones developing tools to control the problems of subjectivity. British civil society shared with its natural philosophers an interest in universalism, namely, organization of its various constituencies into a national system of government, economics, faith, education, and so on. In this chapter, I examine how exchanging provincialism, or idiosyncratic subjectivity, for universalism occupied both early industrial Britain generally and its scientific communities specifically. While the drive to standardize local and subjective knowledge indicated reformist sympathies among natural philosophers, many saw the continuation of hierarchical systems as the best structure for science and society. While some radicals preached democracy, materialism, and deism, most leading natural philosophers walked a more moderate line: the elite still guided the masses, the immaterial mind still directed the body, and God still governed the earth, if only as an absentee landlord. Natural philosophers, with their professed ability to transcend physiological frailty, offered themselves as guides toward a Britain governed by rationally reformed hierarchies.

This chapter examines how natural philosophers prescribed hierarchical divisions of labor along lines of education, class, distance from London, and gender. These hierarchical systems allowed some social and scientific mobility to women, the working classes, rural folk, and nonconformists of merit, while also keeping their subjectivities constrained. We will see in the case studies that follow in Part II that natural philosophers took the same techniques that they used to discipline their own nervously disordered bodies and applied them to the discipline of the sciences' and the state's rank and file. If natural philosophy would not or could not reproduce absolutist order in a reforming age, it would instead promise a well-managed system that specialized in minimizing error and maximizing efficiency. Natural philosophy thus strove to make itself useful to a political economy undergoing dramatic changes. In the aftermath of the French Revolution and the severe growing pains of an industrial economy plagued by bad harvests, an alarmed British elite needed tools to soothe a "nervous" populace. Natural philosophers eagerly cultivated these tools to address such issues and to secure their position and even their own bodies.

Inductive Hierarchies

Half a century ago, Edwin Boring described what he deemed the two major competing methodologies in the history of psychology: "Where phenomenology is egoistic, asserting the validity of individual observation and insight, experimentalism is diffident, mistrusting individual observation and relying upon controls, procedures without knowledge, and the other techniques that have been devised to achieve assurance in the face of the unreliability of human observation."[3] In practice, natural philosophers found themselves negotiating a middle path between these two impossible extremes. The ideals that Boring described sound quite neat, but the practicing scientist has found it difficult to either flatly ignore or disappear into his own ego. For instance, while the Baconian rhetoric that peppered nineteenth-century scientific texts sounded radically empiricist, it also papered over a widespread methodological shift in British science toward analogical, hypothetical, and deductive thinking.[4]

At least one remnant of Baconian philosophy persisted at the end of the Enlightenment. The collection of facts was considered a task achievable by virtually anyone. The question remained, however, whether such democratic impulses should extend to the sciences' more complex tasks—induction, deduction, hypothesizing, analogical reasoning. Bacon certainly had not thought so. And neither did Scottish philosopher William Hamilton, whose evaluation of the work of physician-chemist William Cullen is tellingly blunt:

> In physical science the discovery of new facts is open to every blockhead with patience, manual dexterity, and acute senses; it is less effectually promoted by genius than by cooperation, and more frequently the result of accident than of de-

sign. But what Cullen did, it required individual ability to do. It required, in its highest intensity, the highest faculty of mind—that of tracing the analogy of unconnected observations, of evolving from the multitude of particular facts a common principle, the detection of which might recall them from confusion to system, from incomprehensibility to science.[5]

If any "blockhead" could discover new facts, then perhaps that task should be left to the blockheads. So went one argument for the division of labor, which advocated the rational separation and assignment of atomistic tasks within the factory, laboratory, and other sectors of the industrial economy. The level of skill for most tasks would thereby be minimized, and efficiency maximized.

Once Adam Smith outlined the principle for the division of physical labor, the next generation of natural philosophers such as Gaspard de Prony and Charles Babbage argued that the same rules could—and should—apply to mental labor. De Prony realized the idea when producing mathematical tables for the French government on the application of the new decimal system. He divided his workers into sections, each of which performed specific tasks of varying levels of skill, from simple arithmetic up to writing the analytical formulas. Babbage made this idea the foundation of his difference engine, hoping to make a machine perform natural philosophy's blockheaded, but necessary, tasks. Relieved of the "intolerable labour and fatiguing monotony of a continued repetition of similar arithmetical calculations," natural philosophers could concentrate their full attention on theory and advanced mathematics.[6] Taking de Prony's and Babbage's lesson to heart, some inventors—such as one exhibitor at the 1855 Paris Universal Exhibition of Industry—promised that their devices would spare "all the fatigue resulting from the great attention necessarily kept up in calculations of any importance."[7] Such devices supported a division of mental labor between machines (which performed the simplest functions), various levels of less-educated people (who shouldered the next-most complicated tasks), and the well-educated (whose time was freed for writing theories, designing models, and overseeing operations). Dugald Stewart taught his philosophy students at University of Edinburgh that while division of labor made most physical and simple intellectual tasks more efficient, the economy would always need a class of creative individuals whose labor was not so circumscribed. Inventions, he said, came from the mind of the employer or the "speculative observer," not the workman with his narrowly focused attention.[8]

The division of labor proved a malleable tool for justifying all manner of social inequalities. For example, Peter Mark Roget used the principle to explain

human superiority over animals. While the bodies of "lower" animals performed their functions with "the smallest number of organs," he wrote, in
"higher" animals "the separation of offices becomes . . . more complete."[9] A
similar logic shaped how European anatomists understood the different races'
mental sophistication. European visitors to the Cape of Good Hope, for example, remarked on the superior visual skills of the Bushmen (San). The San's
ability to see clearly at great distances became both an object of envy and denigrated as mere instinct. Europeans purportedly divided their mental labors
more finely among operations such as imagination, will, and reason, but one
cannot help but notice the envy that many European naturalists felt for
Africans' "instinctive" visual sense.[10] Once again we see how uncertain was the
confidence of early industrial social hierarchy: even a principle as powerful as
the division of labor entailed an anxiety about the European intelligentsia's
physiological superiority.

However shaky the foundations, the new principles of political economy
were employed to justify what I will call an *inductive hierarchy,* in which machines and less trustworthy individuals produced the simplest goods (facts),
whose manufacture and synthesis were managed by an elite class of natural
philosophers.[11] This inductive hierarchy manifested itself in a number of analogous ways. For example, we can also find an inductive hierarchy undergirding
mind-body dualism. If the body performed simple, literally mindless tasks, the
mind directed operations and possessed humanity's higher faculties. Likewise,
inductive hierarchy also structured a popular view of Christian faith.[12] God
oversaw the universe just as the mind oversaw the body, just as the manager
oversaw production, just as the natural philosopher oversaw fact collecting. In
all of these hierarchies, the higher authority (God, the mind, the factory manager, the philosopher) transcended simple mechanical and inductive laws. The
philosopher had access to deduction; the manager could make independent decisions about factory operations; the mind could not be explained according to
the physiological observations about the brain and nervous system; and hu-
mankind could only ever hope to approximate God's true laws.[13] In numerous
scenarios in industrial society, then, we find intelligence *managing,* automatons
doing—all in the service of economy and morality.

One's position in the hierarchy did not need to be static. Even the lowly
factory operative might undertake some managerial responsibility over machines, as economist David Ricardo explained: "There is no new creation of
machinery which entirely supersedes the use of the labour of man—he must
regulate its motion and velocity—he must procure coals for the fire necessary

to work it—he must attend to its annual repairs."[14] Though Ricardo would later change his mind about this, whigs such as Richard Jones and Babbage faithfully reasserted the coexistence of mechanization and minding:

> Where one portion of the workman's labour consists in the exertion of mere physical force, [Babbage wrote,] as in weaving and in many similar arts, it will soon occur to the manufacturer, that if that part were executed by a steam-engine, the same man might, in the case of weaving, attend to two or more looms at once; and, since we already suppose that one or more operative engineers have been employed, the number of his looms may be so arranged that their time shall be fully occupied in keeping the steam-engine and the looms in order. One of the first results will be, that the looms can be driven by the engine nearly twice as fast as before: and as each man, when relieved from bodily labour, can attend to two looms, one workman can now make almost as much cloth as four.[15]

Astronomer Royal George Airy likewise lauded those among his computers who sought to "distinguish themselves as observers," a higher class at the Greenwich Observatory.[16] Through initiatives like the Mechanics' Institutes, reformist men and women of science figured that the working classes could eventually leave the manual labor to the machines and instruct themselves in the life of the mind. Engineer James Martineau told the Liverpool Mechanics' Institute that "that natural machine, the human body" had been "depreciated in the market," and warned that "the *mind* must get into business without delay."[17] John Herschel agreed, and knew that to become a pillar in the scientific community, he had to do more than observe. "The credit of a mere observer is not exactly what I am at," he told his friend William Whewell soon after graduating from Cambridge. "Priority of observation has in itself no merit, except as an evidence of diligent enquiry."[18] Educating the working classes to observe certainly constituted an improvement, but the greater glory of discovery awaited the well-educated natural philosopher.

We find an especially rich example of this inductive hierarchy at work in the career of William Rowan Hamilton. Though grateful for his 1827 appointment as professor of astronomy at the University of Dublin and Royal Astronomer of Ireland, he feared that observational duties would interfere with his true love: mathematics. He also found that too much observing adversely affected his health. His mentor, Thomas Romney Robinson, offered to teach Hamilton enough observational skill that he would not embarrass himself in front of his assistant. But otherwise, Robinson advised, Hamilton need not feel shame in his lack of zeal for observation. He had good company in Johann Franz Encke, George Airy, John Pond, Heinrich Schumacher, Friedrich Bessel—all the best in the business. Hamilton therefore gladly handed over much of the observing and calculating to others, even imploring several of his sisters to take over some of his telescope work. His eldest sister, Grace, finally consented for a time to do many of his observations and calculations, and to fortify her brother's scientific quest with her "female sympathy." Hamilton then happily pursued a brilliant career in mathematics.[19]

The argument I am making here—that hierarchical relations buttressed the sciences just as much as other areas of industrial society—might seem to contradict an important historical literature on mechanization, standardization, and objectivity in the nineteenth-century sciences. Lorraine Daston and Ted Porter, among others, have demonstrated the widespread desire among nineteenth-century scientists to "de-skill" their work in order to guarantee the objectivity of the knowledge that they produced. Indeed, devices as diverse as the photographic camera, the method of least squares, and the Royal Greenwich Observatory's standard forms were designed to reduce the extent to which subjective perceptions or theoretical prejudice entered scientific observation. We should recognize, though, that claims to totally mechanical objectivity were *rhetorical strategies* or *goals*. In reality, the very same natural philosophers who argued against a scientific aristocracy and for the democratization of skill, in a pinch reserved for themselves a privileged status in the scientific community. As Richard Yeo phrased this idea, "There was . . . a danger that this emphasis on the accessibility of inductive method could be construed to imply that scientific method was simply a matter of common sense."[20] To the minds of most natural philosophers in the first half of the nineteenth century, governance of the sciences required broader participation and social mobility. However, scientific practice could never become entirely mechanized and democratic. The reforming vision of early industrial natural philosophy, like the reforming vision of early industrial government, required intelligent authority that was irreducible to factory cogs.

The reassertion of this hierarchy often seemed like the only way out of epistemological jams. The foot soldiers of inductivism—the body, the observers—did not always behave as they should. They sometimes provided inaccurate pictures of nature. Furthermore, the automatic controls designed to eradicate such errors rarely worked perfectly. So, while in a perfect world managers would not be necessary, in the imperfect world below heaven real science required them.

Ideally, these managers attained their position through their meritorious reason and judgment, not simply through their social privilege. The reformers who dominated early industrial natural philosophy may not have been radical democrats, but neither did they approve of unchecked aristocratic governance. As these reformers aged, confronting more and more frequent physical, mental, and managerial problems, their belief in the need for a managerial class often grew from a stopgap into a full-blown philosophy.[21] This might seem like an overgeneralization. How, after all, should we handle men such as William Whewell whose a prioristic and elitist views contrasted so sharply with those of a Babbage or a Brewster, who sought to open the field of science through organizations like the British Association for the Advancement of Science? Despite these and other differences, a common thread remained: virtually every prominent natural philosopher in this period explicitly or implicitly preserved some kind of inductive hierarchy. They might have wished to open the upper ranks to all potential comers, but in the practice of science, *someone* had to arbitrate.[22]

Still, we should remember that belief in an inductive hierarchy was a privileged perspective. For example, Anne Secord has argued that despite their best efforts, elite botanists could never fully absorb the work and skill of local artisans into one totalizing system. We therefore cannot swallow uncritically the metropolitan mind-set that elite knowledge equaled truth or that that mindset went unchallenged in its definition of good local knowledge.[23] The fact remains, however, that the subjects of this book—mainly men belonging to an elite, primarily metropolitan class of natural philosophers—were wary of unmanaged local knowledge in its various forms. We will now examine how provincial and (in a later section) female contributors to the sciences seemed to require just such management.

Provincialism as a Kind of Subjectivity

We saw in the first chapter that a kind of "mental provincialism" posed serious challenges to the epistemology and social structure of early industrial natural philosophy. Innate mental differences, different life experiences, and the will all had the power to construct an individualized psychology. London physician Henry Holland considered this the greatest problem facing medical epistemology: "Idiosyncrasy, as arising in most cases from unappreciable causes, is the most absolute and inevitable difficulty in medical evidence; since no accumulation of instances, such as might suffice for the removal of all other doubts, can secure us wholly against this source of error."[24] Natural philosophers found themselves continually challenged to explain what universal mental features remained, and how individuals' ideas could be reliably communicated. Likewise, early industrial society—its sciences included—faced the different question of how to weave its local parts into a nation, into a healthy body politic. Natural philosophers considered their disordered bodies and their locally fragmented scientific community and nation as equivalent projects in hierarchically organized reform.

European states experienced significant power struggles between centralized and local government in the eighteenth and nineteenth centuries. In Britain, for instance, in the early modern period, parish and (to a lesser extent) county governments enjoyed substantial independence. Late in the eighteenth century, however, a growing class of reformers argued that centralized governance would better address mounting problems such as poverty, crime, workplace safety, urban contagion, and insanity. In the first few decades of the nineteenth century, a motley crew of Benthamites, rationalistic dissenters, political economists, enlightened industrialists, and others outside the traditional, landed elite formed a loosely woven reform movement that would achieve substantive change through the 1832 Reform Act and numerous other pieces of legislation that both broadened and re-stratified Britain's political and economic landscape. These same groups simultaneously became more vocal within British natural philosophy.[25]

In an important reassessment of British class formation, Dror Wahrman has argued that the "middling sorts" in the eighteenth century divided roughly into two camps: those who defined themselves according to a cosmopolitan, "national" culture, previously identified exclusively with the aristocracy; and those who supported "communal" or "provincial" cultures.

> For the first time in their collective experience, the "middling sorts"—urban and rural—had a choice: to join in and try to become assimilated to some degree in this novel and attractive culture, a culture which was London-centred and London-oriented; or to assert their distinct values and culture, focused on their local community, against this alien intrusion. The two alternatives from which to choose were particularly accentuated in provincial towns, where there was easy access to genteel culture as well as a variety of appealing options for developing local civic pride.

Wahrman makes a compelling case that provincial versus national sympathies divided eighteenth-century British elite culture more cleanly than did class.[26]

As Wahrman admits, however, the distinction between these two groups can be overdrawn. Not every individual identified strongly with one camp, and some synthesized the two approaches. He cites as an example the Manchester Literary and Philosophical Society, and specifically Arnold Thackray's famous reading of it. The society *both* committed itself to polite culture *and* rejected traditional aristocratic values. "It may be," Wahrman says, "that for every person of the 'middling sort' who made a clear choice, there were two who never made up their minds."[27] This analysis helps us understand a new generation of natural philosophers in the early industrial age who valued provincial knowledge and organization at the same time that they sought to organize it under national standards. Wahrman's argument provides a broader framework for the ambivalence that many natural philosophers felt about whether to value unadulterated provincial knowledge (because it increased the inductive pool of facts) or to risk stifling it through the imposition of more systematic procedures, universal laws, and metropolitan supervision.

The loose alliance of reform-minded natural philosophers that I described in the previous chapter—manufacturing, liberal Oxbridge, and newer academic interests with a common desire for a coordinated national science—opted for a combination of these two strategies. As John Herschel put it in his popular *Preliminary Discourse,*

> To avail ourselves as far as possible of the advantages which a division of labour may afford for the collection of facts, by the industry and activity which the general diffusion of information, in the present age, brings in to exercise, is an object of great importance. There is scarcely any well-informed person, who, if he has but the will, has not also the power to add something essential to the general stock of knowledge, if he will only observe regularly and methodically some particular class of facts which may most excite his attention, or which his situation may best enable him to study with effect.[28]

Why was the management of provincialism so important to industrial-era science? Because, in part, with industrialization came the rapid transformation of several provincial towns into thriving urban centers. Manchester's population, for instance, increased a whopping tenfold between 1760 and 1840. By virtue of that growth, these cities developed audiences large enough to sustain provincial organizations like the literary and philosophical societies and Mechanics' Institutes.[29] These new organizations frequently established themselves upon principles (such as rational dissent) that were foreign to traditional

London scientific elites. The shifting center of gravity in science disturbed some, as exemplified by Joseph Banks's attempts to block the formation of specialist societies in London.[30] Even among those who agreed on the importance of "diffusing" knowledge across a wider swath of the population, the proper method of doing so instigated much debate.

This debate focused on the relationship between science done provincially and science done nationally or in the metropolis. Local information might simply serve local interests, as in the case of the Geological and Polytechnic Society in Yorkshire, which gathered information about coal for the benefit of the mining industrialists who were its members.[31] Some wondered if these local efforts might not be more usefully directed toward national, even imperial, surveys.[32] The organization of magnetic and meteorological stations and the Geological Surveys are just two of the most famous examples of this idea in action. Such projects garnered support from the same gentlemen of science who sought to reform the highly traditional Royal Society and to support the new British Association for the Advancement of Science in the 1830s (including Edward Sabine, Humphrey Lloyd, Francis Baily, Francis Beaufort, and John Herschel).[33] The systematized procedures used in these surveys were meant to guarantee the reliability and equivalency of each observer's and station's data—and to allow a small, centralized cadre of natural philosophers to convert those data into law. In much the same way, Edwin Chadwick's public health surveys in the provinces enabled the parliamentary reforms of the 1830s–'40s.[34]

In these and other instances, the goal was to encourage provincial efforts to the extent that they became communicable across provincial boundaries. Humphry Davy, who served as president of the Royal Society in the 1820s, expressed this view when he marveled at "the wonderful difference in the nature of men, between those who are insignificant in their powers, and apparently isolated in their influences—who live only whilst they move, and cease to act as soon as they cease to exist; and those whose agency extends over the whole social world."[35] This desire for a coordination of talent and a unity of method in the sciences was more of a touted ideal than a reality. Still, the drive to standardize methods and organize large-scale projects did allow the controlled introduction of a nearly nonindividualized work force in the sciences.[36]

The early history of the British Association for the Advancement of Science vividly illustrates the tensions between provincial and universal objectives within Britain's scientific communities. From its founding in 1831, the association convened each year in a different city outside London. Provincial societies and audiences provided much of the funding, planning, agenda, and presence for the meetings. Unlike the Mechanics' Institutes or the Society for the Diffusion of Useful Knowledge, however, the British Association did not consider popularization its main purpose. Rather, any local work presented at the meetings earned value by being absorbed into a common store of knowledge validated by universal scientific methods. As with political reform, elite scientific reformers sought to increase the political base in order to strengthen national governance, not encourage the fracturing of provincial fiefdoms.[37] To Charles Babbage, who played a foundational role in the British Association, this nationalization effort required not only state support, but that of the public as well. Without the guiding hand of public opinion, he wrote, even an administration eager to support science "would run the risk of acting like the blind man recently couched, who, having no mode of estimating degrees of distance, mistook the nearest and most insignificant for the largest objects in nature: it becomes, therefore, doubly important, that the man of science should mix with the world."[38] The state needed local eyes in order to see, but it claimed to provide the mentality to direct that gaze most beneficially.

Pronouncements such as Babbage's sounded radical in the early nineteenth century, less so by midcentury. As opposed to their savant forebears, a new breed of analytical natural philosophers in the nineteenth century sought to take the best of marginal, manual, and provincial knowledge and organize it nationally under universal systems such as political economy, calculus, and statistics.[39] Their efforts met an ambivalent response from the expanding world of provincial scientific culture. Especially in industrial towns, scientific groups tended to promote the ambitions of "marginalized men" like the professional middle classes and Dissenters, whose politics gradually grew more conservative and mainstream toward the middle of the nineteenth century. Though it did not play out in the same way in every province, the model Arnold Thackray traced for Manchester had its rough analogs across Great Britain.[40] That city's Literary and Philosophical Society relied heavily on Unitarian support, and in fact held its meetings in the Unitarian Cross Street Chapel until 1799.[41] The "Lit and Phil" and other local organizations remained in conflict over whether Manchester culture should strive for independence. Through their "great accumulation of wealth and intelligence," one local leader argued, northern and midland towns could rival London's intellectual life.[42]

In contrast to the prevalent radical dissent among intellectuals in manufacturing towns, commercial towns like Bristol, Bath, Liverpool, and York tended to cultivate conservative learned societies. Far from being "marginalized men," Bristol elites, for example, took their cues from Oxford and Cambridge rather than industry.[43] Similarly, Edinburgh scientific culture originally centered on the medical and agricultural concerns of its local elites. From the eighteenth century onward, the Scottish capital found itself in a variable relationship with national and London science. It achieved a great deal of independence and power through the university's superior medical school, but also watched many of its graduates migrate south. Likewise, Scottish men and women of science sought audiences for their work outside of Edinburgh through vehicles such as the *Edinburgh Review* and *North British Review*.[44] By the 1820s and 1830s, a profound tension had formed between those groups that "had a *national* (or international) cultural horizon" and "those whose cultural interests were almost purely *local*."[45] The University of Edinburgh and its medical school, however, eventually allied itself with the former camp and its desire to "civilize" Scotland. One of the most important achievements in this period was a loose alliance between manufacturing, liberal Oxford and Cambridge, and newer academic interests (like those at University of Edinburgh). The alliance shared a common interest in a coordinated national science. In Wahrman's tug-of-war between national and provincial elites, national interests grew stronger in natural philosophy by midcentury.

All problems of provincialism had a common solution, however tenuous and contested that solution was. The natural philosophers I am concerned with here, who generally lived in metropolitan areas, tended to value local knowledge insofar as it produced facts whose reliability could be trusted, and which could be subsumed within a larger, universalizing, theoretical framework. Subjective, provincial knowledge gained the lion's share of its prestige at metropolitan meetings, in nationally circulated publications, and in universal laws. Local knowledge, in other words, should not be *stubbornly* local.[46] Local knowledge gave the nation its eyes but not its will to see the Truth.

Masculinity and Scientific Work

The attitude that local, embodied knowledge required careful supervision was not only mapped out according to metropolis and province; it was also gendered. To many a scholar in the early industrial age, the active life of managing, deducing, and experimenting seemed more masculine than the passive qualities of sensibility. In fact, it was during this period that nervous sensibility became widely reconstructed as feminine, leaving men to forge a new set of ideal qualities for themselves.[47] A new masculine ideal of scientific work elevated active qualities such as the reason, judgment, will, and attention as principles that would guide perception. Coleridge, for instance, doubted that "any thing manly ... [could] proceed from those, who for law and light would substitute shapeless feelings, sentiments, impulses, which as far as they differ from the vital workings in the brute animals owe the difference to their former connexion with the proper virtues of humanity."[48] Masculinity increasingly became associated with the active powers and with those in whom active powers predominated.[49]

We can witness the difference between the early Enlightenment male's "balanced" sensibility and the industrial male's preference for "active" mental qualities by comparing texts from these two periods. For Enlightenment sensibilities we can look to two popular London periodicals, the *Ladies' Diary* and the *Gentleman's Diary*, both of which provided leisure reading on the arts and sciences. Mathematical problems and riddles filled much of the *Ladies' Diary*, and solving them promised to endow the magazine's readers with "VIRTUE and SENSE with FEMALE-SOFTNESS join'd."[50] Answers to the riddles sent in by a core group of educated men (mostly) provided the guidance needed to achieve enlightenment. One question sent to the *Ladies' Journal* by "Mira" asked if a "person of sensibility and virtue" could fall in love again after the death of one's true love. In one of his few stabs at a moral query, the later-famous chemist John Dalton replied that "in sustaining the disappointments incident to life, true fortitude would guard us from the extremes of insuperable melancholy and stoic insensibility, both being incompatible with your own happiness and the good of mankind." An appropriate tempering of the passions with reason would enable Mira to endure the shock of losing a loved one "without eventually feeling the mental powers and affections enervated and destroyed by it."[51]

By the mid-nineteenth century, the ideal mental state endorsed by most British natural philosophers had changed. Efficiency seemed more desirable than the balance of reason and the passions. If this meant the neglect of emotion, many natural philosophers willingly made the sacrifice, though often with some regret. In his *Autobiography*, Charles Darwin reified in his own person this maturation from feminine receptivity and balance to masculine control. Losing his appreciation for novels and other aesthetic objects left him sentimental for the "emotional side" of his character, but he considered it a worthwhile sacrifice to attain the patience to experiment meticulously in the service of his "one long argument." Still, he lamented that if he could live his life again he would regain some of his aesthetic sense, "for perhaps the parts of my brain now atrophied could thus have been kept active through use. The loss of these tastes is a loss of happiness, and may possibly be injurious to the intellect, and more probably to the moral character."[52] Darwin's wistfulness for his lost emotion sounds rather like European travelers' envy of the more "primitive" San's eagle-eyed vision. Both losses entailed a mental superiority that was to be relished and a bodily ignorance that was to be regretted.

Michael Faraday, who labored under some of his own physical ailments, argued that these losses could be managed through a better understanding of the self. In one of his popular lectures at the Royal Institution, he lamented how many people allowed their powers of judgment to atrophy. This mental laziness compromised one's character in its fullest sense. Faraday found a cure for this lapse in "*self-education,*" which meant improving one's judgment through the discipline of the imagination, the knowledge of one's abilities and limitations, and the cultivation of "clear and definite language," patience, and humility.[53] Fortunately, he assured his audience, such mental discipline

> requires no unpleasant avowals; appearances are preserved, and vanity remains unhurt; but it is necessary that a man *examine himself,* and *that* not carelessly. On the contrary, as he advances, he should become more and more strict, till he ultimately prove a sharper critic to himself than any one else can be; and he ought to intend this, for, so far as he consciously falls short of it, he acknowledges that others may have reason on their side when they criticise him. . . . *The mind is not enfeebled by this internal admission, but strengthened.* . . . It is right that we should stand by and act on our principles; but not right to hold them in obstinate blindness, or retain them when proved to be erroneous. I remember the time when I believed a spark was produced between voltaic metals as they approached to contact (and the reasons why it might be possible yet remain); but others doubted the fact and denied the proofs, and on re-examination I found reason to admit their corrections were well-founded.[54]

One strove to understand the mind not just to learn facts, but also to improve oneself. Self-examination thereby strengthened rather than diminished the natural philosopher's authority. John Abercrombie concurred that without an investigation of one's intellectual and moral condition, "an individual may make the greatest attainments in science, may learn to measure the earth, and to trace the course of the stars, while he is entirely wanting in that higher department, the knowledge of himself."[55]

Because the demands of natural philosophy could tax the nerves of even great men like Darwin or Newton, many questioned whether the more fragile female nervous system was up to the task. As a wealth of studies on the history of medicine have shown, women were thought to be riddled with physical and mental ailments. The uterus supposedly had a strong link to the nervous system, and thereby could wreak all sorts of pathological havoc—including, most famously, hysteria.[56] Most natural philosophers agreed that women could learn and even teach established scientific knowledge, but feared that the nervous agitation brought on by strenuous knowledge *production* might be too much for them.[57] A typical example was George Airy's rebuff of a young woman who circulated a petition to have Cambridge grant degrees to women. While he believed women capable of appreciating much scientific knowledge, he did not think that "their nature or their employments [would] permit of their mastering the *severe* steps of beginning (and indeed all through) and the *complicated* steps at the end" of a full scientific education. Why they should seek such employment was beyond him, since the "delicacy of a woman's character" seemed "*infinitely* more valuable than any amount of knowledge."[58]

For Francis Bacon, the vulnerability and passivity of women had seemed comparable to the passive accumulation of facts in early science. In its infancy, he had said, a feminine science passively received God's truth, and then later gave birth to the masculine scientist.[59] For many early-industrial natural philosophers this translated into a scientific hierarchy where women and other undereducated participants fulfilled the necessary but lower observational tasks, while the male philosophical masters orchestrated events. However, the gender politics of experimental science were never quite so clear-cut, for the experimenter never altogether ceased to be an observer. In fact, Barbara Stafford argues, "the moment the result of the experiment manifests itself the experimenter as interpreter must vanish, must transform himself instantaneously back into pure observer."[60] In effect, this simply meant that the male experimenter could liberally appropriate desirable feminine qualities and observational techniques, while upward mobility for the uneducated female observer remained barred.

These points are well illustrated in John Herschel's evaluation of Mary Somerville's accomplished 1832 translation of and dissertation on Laplace's *Traité de mécanique céleste.* Undergirding Herschel's review was the popular assumption that even though feminine appreciation of beauty and the sublime should not dominate natural philosophy, it should certainly temper masculine overemphasis on utility:

> The female bosom is true to its impulses, and unwarped in their manifestation by motives which, in their sterner sex, are continually giving a bias to their estimates and conduct. The love of glory, the desire of practical utility, nay, even meaner

and more selfish motives, may lead a man to toil in the pursuit of science, and adopt, without deeply feeling, the language of a disinterested worshipper at that sacred shrine—but we can conceive no motive, save immediate enjoyment of the kind so well described [by Somerville] . . . which can induce a woman, especially an elegant and accomplished one, to undergo the severe and arduous mental exertion indispensable to the acquisition of a really profound knowledge of the higher analysis and its abstruser applications.[61]

In a fascinating twist, Herschel here co-opted the female bosom and its abstracted good qualities for natural philosophy, while wondering at the scientific participation of actual, frail-bodied women.

Far from wincing at Herschel's ambivalence toward her femininity, Somerville basked in his assessment. Before sending her important dissertation on Pierre-Simon Laplace to the printer, she gave Herschel a draft for his approval, "without which I should be unwilling to let it appear before the public."[62] In fact, she continued to send him prepublication drafts for the rest of her career. With some criticisms, Herschel always *did* approve of her work. He greatly admired her intellect, and his correspondence with her contained as much technical detail as his scientific exchanges with men. He even publicly defended Somerville's experimental abilities in the aforementioned review.[63]

However, it is extremely interesting to note *how* it was that this woman attained the status of natural philosopher in Herschel's eyes: "The same simplicity of character and conduct [that characterizes Somerville's work], the same entire absence of anything like female vanity or affectation, pervades the whole of the present work. In the pursuit of her object, and in the natural and commendable wish to embody her acquired knowledge in an useful and instructive form for others, she seems entirely to have lost sight of herself."[64] In other words, to move beyond the normal feminine role as observer, and to establish herself as a reliable experimenter and commentator on cutting-edge physical science, Somerville had to shed her feminine qualities and lodge her intellectual capabilities in another "body." The finished work thereby supposedly left no trace of its feminine origins. That Somerville could achieve even this was a credit to her fine breeding and education, for Herschel found nothing to "betray a latent consciousness of superiority to the less-gifted of her sex, or a claim either on the admiration or forbearance of ours, beyond what the fair merits of the work itself may justly entitle it to. There is not only good taste, but excellent good sense in this."[65] Somerville's class standing enabled her to abstract her knowledge from the worst vagaries of her sex.

Herschel's attitude was by no means unique. Somerville also was admired by her daughter and biographer, Martha, and the popular science writer Jane Marcet for uniting the "talents and acquirements of masculine magnitude" with the "most sensitive and retiring modesty of the female sex."[66] We must also note that Herschel was asking no more of Somerville than he asked of himself. Namely, he expected women *and men* to strive constantly to discipline their bodies. Without the training in reason and judgment that respectable men normally received, women simply had a harder time of it. Somerville agreed. "I have always considered a highly-educated aristocracy essential," she reflected, "not only for government, but for the refinement of a people."[67]

Somerville purchased her tenuous place in the scientific elite at a price— she had succeeded *despite* her womanly embodiment. At her greatest moments, she had "lost sight of herself." In similar fashion, Mary Wollstonecraft in her *Vindication of the Rights of Woman* praised Catherine Macauley and other women authors for a style in which "no sex appears, for it is like the sense it conveys, strong and clear."[68] Sixty years later, James Clerk Maxwell echoed this sentiment when he argued that objectivity required one to have an active, public life. Women who turned inward and reflected on their "thoughts, motives, and secret sins" were "caged in and compelled to criticise one another till nothing is left, and you exclaim:—'Madam! if I know your sex, By the fashion of your bones.' No wonder people get hypochondriac if their souls are made to go through manoeuvres before a mirror."[69] Wollstonecraft, Herschel, and Maxwell all agreed that too much femininity resigned one to a life of passive observation and cannibalistic introspection.

There was no better object lesson in this than Herschel's own accomplished aunt Caroline, who was made an honorary member of the Royal Society the same year as Somerville, but who still hesitated to accept the Royal Medal that society wished to present to her, because she knew "how dangerous it is for women to draw too much notice on themselves." She saw herself as a "mere tool" that her brother "had the trouble of sharpening and to adabt *[sic]* for the purpos *[sic]* he wanted it, for Lack of a better." Ever cognizant of the constraints imposed by her womanhood, Caroline never trusted herself as an "active observer"—she barely had confidence in herself as a *passive* observer, preferring to fashion herself as a "pointer dog" to her brother William's astronomical hunter. In her brother's and nephew's hands, raw astronomical data became theoretically constructed heavens and experimentally analyzed hard bodies. In the minds of all three Herschels, Caroline could contribute simple, passive observation and little more.[70] If this pleased her contemporaries, two generations later it mortified the American astronomer Maria Mitchell. While Caroline Herschel's scientific achievements could inspire young women, "what she did not do is a warning. Has any being a right not to be? When Caroline Herschel so devoted herself to her brother that on his death her own self died, and her life became comparatively useless, she did, unconsciously, a wrong, and she made the great mistake of her life."[71]

Caroline Herschel's example quite vividly reminds us to take notice of the gendered nature of discourse about minds and bodies in natural philosophy. For her and other women, the ability to exercise reason and willpower over their "natural" sensibility did exist. But the natural philosopher in the industrial age believed that men achieved such things more readily than women. The ability to turn abnormal physiology and subjective experience toward general knowledge thus stood as a masculine trait, if not a trait enjoyed exclusively by men.

We have seen that natural philosophy carefully maintained an uneasy peace with its dependence on subjective knowers such as provincials and women. Awkwardly, the production of universal laws was forever locked in a relationship with the production of local facts by local knowers. This left the elite natural philosopher never entirely certain that his universal law described collective experience. Surely masculine, metropolitan, and mental will provided the management needed to guide such sophisticated production. One key problem remained: How could this will be communicated to working hands and eyes?

The Problem of Communication

The natural sciences had developed a number of languages to handle problems of subjectivity, to create lines of communication across borders. These languages included technical vocabularies, rationalized nomenclatures, mathematics, training, and machines. For example, we can understand Babbage's difference engine as a hardware language for expressing his conception of mental operations.[72] We might also view the efforts of Cambridge's Analytical Society as an attempt to replace the uncommunicative features of Newton's fluxional calculus.[73]

The problem of communication stood at the heart of concerns about subjectivity, provincialism, and science. Provincialism implied subjectivity. Much of the problem was intrinsic to language itself. For instance, one of the fundamental conveniences of language was that there was no one-to-one correspondence between word and idea. In *Gulliver's Travels*, Jonathan Swift pointedly demonstrated how cumbersome such a strict correspondence would be.[74]

Swift's satire aside, one was left with the problem that language, in all its complexity, upset what might have been a clear relationship between the provincial and the universal, for if on the one hand language made public knowledge possible, it also might jam an interpretive wedge between nature and the mind, and between knowers. Replication posed a particularly serious problem: for example, how could one accurately reproduce the experience of someone with a nervous disorder? With limited possibilities for replication, how could one trust that a person's reporting of their experience was accurate? Communication was, and is, thus a fault line running through the bedrock of objectivity.

Self-consciousness prevented that fault line from splitting scientific authority asunder. If one knew one's own interpretive quirks, one could eliminate them from — or at least minimize them in — the final public account.[75] Indeed, the "very act of verbalizing experience implies the public constitution of knowledge." As soon as an individual uses language, a social institution, to describe her experience, it ceases to exist solely as an individual experience.[76] Observational and experimental reports require significant effort, then, because the author must find the right words to convey the fullest meaning of one's experience. Even those with normal nervous systems found that communication always had to act as mediator for the psychological and phenomenological effects of subjective experience. David Gooding has argued that scientists always reach agreement about the meaning of new observations through a process of "construal," or a description of the practices through which a novel phenomenon is observed. Only over time and through negotiation does the construal disappear and interpretation of the phenomenon become self-evident.[77]

In other words, through study of the nervous system natural philosophers learned that communication required negotiation. If words had more than one meaning, if more than one description could relate an experience, then arriving at the truth required natural philosophers to negotiate what would stand as an authoritative description of a phenomenon. One way to do this was to define terms more precisely. For instance, we might agree that when we describe something as "red" we only mean those frequencies of light that correspond to the red part of the spectrum. Another technique for negotiating language is qualitative precision. Rather than use qualitative language like "red," we could agree to match each individual sensation to "red #1," or "red #2," and so on, on a standard color chart. Both methods gained favor in the eighteenth and nineteenth centuries.

Natural philosophers developed such tools because they had long considered language and perception untrustworthy. Under the Lockean empiricism that dominated British natural philosophy from the Enlightenment, both language and perception were systems of conventional signs with no organic relationship to real objects. Rather than resigning natural knowledge to mere phenomenology, however, Locke argued that philosophy could approach truth by disciplining its discourse. As in Thomas Sprat's *History of the Royal Society,* plain language (and, one might argue, plain sensations) would minimize the rhetorical flourishes that obscured true knowledge. Unique experiences *had* to be communicated through the persistently awkward medium of language, appropriate as a tool at least because (as the empiricists would have it) it was an acquired faculty just as perception was. But further comparisons of singular experiences might lead to a reasonably accurate nosology. What had once been ineffable self-knowledge had the potential to become objective knowledge through the intervention of tools such as prisms, carefully executed experimental trials, and technical language.[78] Along with the new biblical critics, natural philosophers and physicians configured language and even sensations as signs. Proper interpretation was no longer inherent to the signs themselves, but rather came with experience and training.[79]

Besides various "objective" protocols, an observer's training, education, and standing ensured that one could reliably translate personal experience into reliable knowledge. Elie Wartmann, natural philosophy professor at the Academy of Lausanne, figured that the "vast extent of . . . [color-blind chemist John] Dalton's knowledge is a guarantee of the fidelity and care he has exercised in analysing his [color] sensations, and renders the sketches of them which he has published very valuable."[80] David Brewster likewise trusted the self-reporting of two color-blind observers enough to print their accounts verbatim in one of his reviews. He credited these two observers, Frederic Dyster and an unnamed colonel, as "gentlemen who are well qualified to describe their sensations. Both of them are of middle age, in good health, and possessing excellent vision." Dyster claimed to have developed a "very acute perception of shade" which compensated in many ways for his insensitivity to color. This extra sense ensured the cultivation appropriate to his rank. In his color blindness Dyster was quite unlike "the person without ear . . . in regard to noise," for he fully appreciated fine art and the "gorgeous displays of polarization." The colonel also reported that he had "in every other respect a peculiarly good sight" (see chapter 3).[81]

Communication could even pose problems for those with normal vision. Helmholtz feared that sensory reports about the external world corresponded "only in some such way as written characters or articulate words to the things they denote. They give us, it is true, information respecting the properties of things without us, but no better information than we give a blind man about colour by verbal descriptions."[82] That a devoted empiricist such as Helmholtz could express such doubt about the reliability of the senses and language is not altogether surprising when we remember the importance that he attached to instruments and other conventions of experimental procedure. As Helmholtz knew well, induction was never as simple as some made it seem, and required more than a litany of simple sensations and facts strung together with simple words.

Helmholtz's concerns bring us back to the heart of this book. Natural philosophers not only worried about the reliability of information communicated from subjective knowers such as workers, provincials, and women. They also wondered about the reliability of information communicated from their own subjective bodies, whose workings could be just as mysterious. We have seen in this chapter that natural philosophers used multiple strategies to manage all these variants of subjectivity. These strategies grew from the same tree as the broad array of political, social, economic, and religious reformers, developed rational systems which they hoped would organize and tame the various peculiarities of British experience.

This chapter's collective picture depicts the idealized natural philosopher managing an array of machines and less-educated workers in a well-oiled drive for reliable knowledge. That idealized managerial figure was male, well born, well educated, disciplined, and either lived in London or at least spent significant time there. Natural philosophers clung to this ideal like a life line against the various problems of doing real scientific work. One of those problems was ensuring that the facts were transparently communicated from province to center, from worker to manager, from female and working-class observer to educated, male theorist. As we will see in the following three case studies, natural philosophers found that they had to get their own bodies in working order before they could hope to manage scientific communication in the far larger body politic. Cleverly, natural philosophers learned to package their nervous struggles as a kind of pilgrim's progress that imbued them with more, not less, authority to manage their nation's problems. If early industrial natural philosophers lacked the Enlightenment's confidence in certainty, they recuperated themselves with a tonic brewed in a reform-minded, nationalist, industrial mentality.

PART II

The Nervous Conditions

THREE

Provincialism and Color Blindness

Methinks I see him, how his eye-balls rolled
Beneath his ample brow, in darkness paired,
But each instinct with spirit; and the frame
Of the whole countenance alive with thought,
Fancy, and understanding; while the voice
Discoursed of natural or moral truth,
With eloquence and such authentic power,
That, in his presence, humble knowledge stood
Abashed, and tender pity overawed.

—William Wordsworth

JOHN GOUGH'S CAMEO APPEARANCE in Wordsworth's "The Excursion" (above)[1] marked his status as a minor legend in the Georgian period. A childhood bout with smallpox left Gough blind. Undeterred, he embarked on a career as a mathematics tutor and botanist. He could distinguish plants according to their tastes and smells, and even said he could taste color. While Wordsworth used Gough's example to illustrate the power of "thought, / Fancy, and understanding" to rise above physical limitations, the blind man's skill also extended to those pursuits thought to rely almost solely on the power of sight. David Brewster marveled that Gough could apply his mind so fruitfully to all manner of highly visual sciences, including astronomy, chemistry, medicine, optics—even the nature of vision.[2] Gough's former pupil, John Dalton, admired "what genius, united with perseverance and every other subsidiary aid, can accomplish, when deprived of what we usually reckon the most valuable sense."[3] In the Victorian period, after his death, fascination with

Gough peaked thanks partly to the success of two of his pupils, Dalton and William Whewell. That two such acutely observant natural philosophers could have received their earliest education from a man without the faculty of sight, suggested how far genius could transcend physical debility.[4]

Among the lessons John Dalton must have learned from his early patron was this power of transcendence. In fact, in many ways Dalton succeeded even more at this feat than Gough did. Granted, Dalton's problem was less serious. His color blindness simply limited the hues he could see to yellows and blues. Still, the limitation was potentially damaging enough for a young natural philosopher interested in chemistry. The fact that Dalton became lionized in nineteenth-century Britain as the father of modern chemistry signified his victory over not only his physical circumstances, but also his humble, Quaker, provincial upbringing. In fact, I argue in this chapter that early-nineteenth-century reformers closely associated all of these potential limitations (provincialism, Quakerism, and color blindness). Reformers in the British scientific community forged this association as a way of both understanding Dalton's particular case and addressing the problem of subjective knowledge described in the previous chapter. Dalton served as an exemplar in early industrial Britain of how to use the tools of natural philosophy to manage subjectivity.

One reason that color blindness captured natural philosophers' interest was that several prominent colleagues besides Dalton shared the condition, to include moral philosopher Dugald Stewart, instrument maker Edward Troughton, and civil engineer William Pole.[5] To understand what this vision problem meant to that community, however, we first need to take a broader view of the connections between color blindness and provincialism in British culture, and why these two issues posed particularly thorny problems for natural philosophers.

Color blindness served as an important site for clarifying how to obtain reliable, local, empirical data and how to organize these local facts into universal scientific knowledge. As explained in the previous chapter, during the early industrial age provincial towns began to challenge traditionally dominant London culture. In response to this pressure, two social groups pushed to reconvene centralized power: metropolitans jealously guarding their former political and cultural monopoly, and liberal and moderate reformers seeking to standardize and centralize governance through such diverse mechanisms as the census, sanitary commissions, and national scientific surveys. Natural philosophers, and particularly those interested in color blindness, generally fell into the latter camp. They saw standardization as a key tool for generalizing subjective experiences of those such as the color-blind or the provincial.

The centrifugal force exerted by the provinces pulled so powerfully that it began to act on issues that we might now think had nothing to do with provincialism. Take, for example, the specific case addressed here: color blindness. Reports on the subject from this period tended to emphasize the epistemological tension between the isolation of subjective experience and the ultimate goal of universal knowledge. How could one be sure that color-blind people accurately reported their inner experiences? How could subjective accounts of color blindness (or even normal sight) ever coagulate into general laws of vision? Men of science also asked this kind of question in larger contexts: Could the modern philosopher still work in solitude on individualized projects, or should collective Baconian schemes become the rule? Could natural philosophy in the provinces ever match the sophistication of metropolitan science? What discourse should the sciences adopt in order to assure a common tongue that could cross parish, even national, boundaries?

COLOR BLINDNESS AS A METAPHOR FOR PROVINCIALISM

Before Dalton's 1794 announcement of his color blindness to his provincial audience, the condition had received little notice. By the time Charles Babbage published his autobiography in 1864, however, the investigation of abnormal vision—and color blindness in particular—had become one cipher for the natural philosopher's mastery of the body and of provincialism. Babbage chose this metaphor to express his admiration for the late Prince Albert's curiosity: "Rarely indeed can some deep-searching mind, after long comparison, perceive the real colours of those translucent shells which encompass kindred spirits; and thus at length enable him to achromatize the medium which surrounds his own. To one who has thus rectified the 'colour-blindness' of his intellectual vision, how deep the sympathy he feels for those still involved in that hopeless obscurity from which he has himself escaped."[6] Babbage especially appreciated Albert's ability to abandon aristocratic prejudice against material progress and become a man of action, as signified by his organization of the 1851 Great Exhibition. Less visionary leaders were "isolated spirit[s]," their sight tinted by their limited circumstances. Only the few could see clearly through the blinding light of progress on display at the exhibition. Physician White Cooper reported that he knew several people who, upon visiting the Great Exhibition, had had their color vision "temporarily blunted by the excitement to which it was exposed in that brilliant scene."[7]

Throughout the early industrial period, color blindness served as a useful weapon in a variety of ideological battles. In the late Enlightenment, one prominent strand of commentary on color blindness bemoaned it as an aesthetic handicap. Having a blunted sense for the subtleties of color or tone did not mix well with life in high society. In one of her early days at court in 1785, Frances Burney conversed with George III about people who had no ear for music. "Lady Bell Finch once told me," he said, "that she had heard there was some difference between a psalm, a minuet, and a country dance, but she declared they all sounded alike to her! There are people who have no eye for difference of colour. The Duke of Marlborough actually cannot tell scarlet from green!" The king then horrified Burney by relating his own occasional mistakes in color judgment. "How unfortunate for true virtuosi," she moaned, "that such an eye should possess objects worthy of the most discerning—the treasures of Blenheim!"[8]

An "educated man" betrayed an equal horror at the coarseness of color blindness when he wrote to Glasgow physician Hugh Colquhoun: "I ought to state at the outset, that I do not invariably confound the colours to the distinction between which my eye is generally insensible. Thus the brilliant reds and greens in the plumage of certain birds, and in some fruits, I can always appreciate with perfect accuracy, whilst to the generality of eyes the difference between such colours and those which I am certain to confound is not perceptible; neither can I myself explain in what the peculiarity consists, which thus enables me to appreciate their quality truly." Though this man modestly proclaimed ignorance as to why he could see through his visual defect, the contemporary reader could have discerned the answer. A well-bred, educated individual with color blindness would perceive colors better than an uneducated man born with perfect sight. Still, the man's insensitivity to color caused him some social embarrassment, as when he failed to notice "the complexion of a foreign lady who had purposely substituted Prussian blue for her rouge."[9]

In a culture that so valued truth to nature as an art form, whatever line existed between fact-mindedness and beauty-mindedness was easily breached. A refined color sensibility was part of an array of skills required both for revealing truth and cultivating taste. "What sense can they [the color-blind] have," one natural philosopher lamented, "of the excellences of Titian, Rubens, Etty, or Lance; or of the painted music of Turner?"[10] Insensitivity to color and provincialism went hand-in-hand. The cultivated person who had to distinguish colors because of the requirements of taste, or the natural philosopher who applied this discrimination to studying nature, possessed the fine-tuned attention to color needed to recognize the differences between others' color vision and their own. Ironically, then, only those who really needed a highly developed color sense could tell when they did not have one.[11]

In 1858, chancellor of the exchequer and future prime minister William Gladstone used this developmental thesis to explain the lack of subtlety in Homer's descriptions of color. Richness of vision and the possibility of detecting color blindness were foreclosed to those early cultures whose "colour-adjectives and colour-descriptions . . . were not only imperfect, but highly ambiguous and confused." The "savage," the inferior animal, and the ancient Greek might have (had) stronger vision for distances and extremes of light and dark, but strength did not mean refinement.[12]

If color blindness became a kind of synecdoche for impaired judgment, then the finest natural philosophers needed to tout their ability to rise above such problems. To call an adversary color-blind in the nineteenth century was to invoke an especially new and vivid metaphor for one-sidedness, illiberality—in a word, provincialism. For example, Dalton's close friend, William Henry, extolled Joseph Priestley as one who shunned such dogmatism. "Though this praise may, doubtless, be conceded to the great majority of experimental philosophers," Henry told the inaugural meeting of the British Association for the Advancement of Science, "yet Dr. Priestley was singularly exempt from that disposition to view phaenomena through a colored medium, which sometimes steals imperceptibly over minds of the greatest general probity."[13] French natural philosopher Elie Wartmann even adapted this argument to explain why John Dalton had been the first to make color blindness a public issue. "Are we to conclude from hence that this imperfection is the lot of modern European nations?" Wartmann asked. "It is infinitely more probable that in other regions and in past ages the spirit of observation less mature has not discovered it."[14] Whereas previous generations of Europeans had been mired in their own provincialism, modern men like Dalton could recognize their own limitations and thereby control them.

Later in this chapter we will see in more detail how Dalton and his public reputation as one who transcended provincialism became so closely associated with color blindness. By the mid-nineteenth century, the strength of that association led to even broader comparisons between illiberal behavior and color blindness. In what might have been the first use of color blindness in reference to racial politics, *Punch* responded to the October 1865 Eyre controversy with uncharacteristic solemnity.[15] An article titled "Last Case of Colour-Blindness" reviled English humanitarians for thinking only of the poor treatment of Jamaican blacks, while ignoring the economic hardship that abolition would cause white colonists: "It is strange that fanaticism blinds men, not otherwise foolish, to the truth, and makes them unaware of the feelings of the great mass of their countrymen."[16]

Imperial and class interests encouraged other invocations of color blindness. Sir J. Gardiner Wilkinson, a member of the Royal Society, asserted that while taste in color was innate, manners could and should be taught to the less fortunate.[17] Similarly, Brewster argued that optics and physiology could be put to the service of class-wide improvement.[18] This "democratic" sense of taste—endorsed by many of the most prominent natural philosophers—emphasized the need to learn the rational principles on which taste rested, thus demystifying access to good manners.[19] Training literally sensitized one's nerves and could thereby improve one's vision. If civilization made one prone to nervous disorders, it at least provided the educational tools to minimize the deficiencies.

The 1898 Cambridge anthropological expedition to the Torres Strait further developed this idea into a Darwinian position that physical and cultural evolution were not simultaneous. Physician and Cambridge psychology lecturer W. H. R. Rivers found that Torres Strait islanders had less acute hearing than Europeans but were less susceptible to visual difficulties such as optical illusion, myopia, astigmatism, and color blindness. Despite this advantage, the islanders' color sense was deemed deficient in its *subtlety*—their physical superiority had not guaranteed them the aesthetic refinement of distinguishing fine levels of color difference. The harsh conditions on the islands, Rivers argued, necessitated sharp vision more than intellectual development. He concluded that environmental conditions contributed to the evolutionary development of color vision. By implication, color blindness afflicted Europeans because the race's superior intellectual development had compromised some basic physical skills.[20]

In a very different context, Charles Darwin himself also used the trope to deride illiberality. Reflecting on his critics in his 1887 autobiography, Darwin acknowledged that they might simply call him blind to the grandeur of God's creation. Though the *Beagle* voyage had impressed him with the "conviction that there is more in man than the mere breath of his body," as an old man he no longer felt the tug of natural theology.

> It may be truly said that I am like a man who has become colour-blind, and the universal belief by men of the existence of redness makes my present loss of perception of not the least value as evidence. This argument would be a valid one if

all men of all races had the same inward conviction of the existence of one God; but we know that this is very far from being the case. Therefore I cannot see that such inward convictions and feelings are of any weight as evidence of what really exists.

Any analogy between color blindness and his own God blindness had no force, Darwin argued, because a significant population insisted that its non-experience of God and the soul was objectively true. In contrast, no color-blind person insisted that red did not exist just because she could not see it. As had been achieved by then with color blindness, Darwin hoped to sift out the true aspects of nature (its sublimity) from what was merely subjective (the feeling that that sublimity implied God's presence and power).[21] By the late nineteenth century, then, men of science used color blindness to assert their authority over a wide swath of Victorian concerns, including empire, religion, and class politics. To achieve this, natural philosophers developed a set of standardized methods to study color blindness over the nineteenth century, and then claimed that the reliability of their techniques gave them special abilities to solve problems of provincialism within and beyond the sciences.

As the above examples indicate, seemingly narrow scientific discussions of color blindness took place within—and fueled—a much broader conversation about who had and should have epistemological authority. The study of color blindness can therefore aid our understanding of the shifting place of local knowledge in natural philosophy and early industrial Britain more generally. The natural philosophers discussed in this chapter interpreted their work on color blindness within this larger context. They sought to generalize the highly particular experience of the color-blind both as a means to discover general laws of vision and to make sense of the rapidly proliferating numbers and kinds of scientific knowers. That several prominent natural philosophers were themselves color-blind made it especially important for the community to decide what precisely this visual abnormality meant. Learning the nature of color blindness would take natural philosophers one important step closer to the larger solution of translating provincial into universal knowledge.

Could the peculiar botanical skills of a blind man from Kendal (or even the collective skills of thousands of such men) ever equal the power of Newton's law of gravitation? Or should these idiosyncratic skills be standardized to meet the expectations of a new political economy? The natural philosophers discussed in this chapter admired men such as Gough and Dalton not because of the peculiarity of their vision or, by analogy, their provincial circumstances, but because of their *transcendence* of these things. They valued provincial knowledge, including the knowledge of the color-blind, but not for its own sake. Rather, the real value in color blindness, and provincial knowledge generally, lay in the opportunity it provided natural philosophers to demonstrate their skills at managing subjectivity.

Color in Natural Philosophy

One reason color blindness became such a crucial issue in the drive to generalize idiosyncratic experience was that discerning color was a fundamental skill in many of the natural sciences. Natural historians required it to classify specimens. The color of a celestial object gave the astronomer clues as to its type. The nature of color also played a central part in optics. In fact, until Thomas Young and Augustin Fresnel redirected attention to the propagation of light in the 1810s, the nature of color had been the dominant issue in optics. With the invention of spectroscopy this interest spread to other physical sciences. And though nineteenth-century chemists increasingly preferred quantitative techniques, the color of a substance still provided a first, qualitative indicator about the result of a reaction.[22]

The trouble was that color — and more particularly, color perception — was unreliable for several reasons. Color changed appearance under varying physical and physiological conditions. Furthermore, opticists disagreed whether the lenses of the human eye always or even ever refracted light correctly.

Physical and Physiological Effects on Color Perception

Changes in the external environment or the observer's body could affect color perception. Isaac Newton had shown this experimentally: he placed a patch of gray powder in direct sunlight and a white card in shadow. He adjusted the light until the powder and card seemed to be the same color. When a friend visited soon after, Newton asked him to stand at a distance and tell him if the two objects were the same shade. "After he had at that distance viewed them well, he answer'd, that they were both good Whites, and that he could not say which was best, nor wherein their Colours differed."[23] Opticists in the eighteenth and nineteenth centuries also were fascinated by afterimages, the images left "floating" in front of the eye after staring at an object long enough to fatigue the retina. For example, an observer who stared for some time at a red card would, upon looking away or closing the eyes, see a floating green card. As Buffon, Goethe, Robert Darwin (Charles's father), and others argued, these effects demonstrated the central role of judgment and fatigue (respectively) in color perception.

The subjectivity of color vision could cause real problems in scientific work. Astronomy handbooks presented students with a litany of warnings about color illusions: the thick British atmosphere refracted light in deceiving ways. Viewing celestial objects against the blue of a day-lit sky—even looking at an object too near to a strongly-colored star—could also lead to errors. At night, the astronomer had to take care to allow his eyes to adjust from the yellow, artificial light of his home to the darkness needed to observe.[24]

The very act of studying optics could lead to idiosyncratic color vision. Several of the most prominent natural philosophers in the early nineteenth century temporarily blinded themselves while studying the sun. David Brewster, Joseph Plateau, and Gustav Fechner each lost their sight for several years. Many more experienced a less drastic effect from this kind of experimentation: they temporarily lost sensitivity to the red end of the spectrum. This happened commonly enough that Charles Dickens incorporated it into his novel *Dombey and Son*. In his first appearance in the story, the character Solomon Gills has blood-shot eyes and a "newly-awakened manner, such as he might have acquired by having stared for three or four days successively through every optical instrument in his shop, and suddenly came back to the world again, to find it green."[25]

In any one of these circumstances, an observer could have difficulty deciphering which color effects resulted from the objects observed and which were caused by the observer's own perceptual mechanism. Newton's famous *experimentum crucis* with the prism encouraged him to abandon an Aristotelian understanding of color as fundamentally physiological and instead posit color as an intrinsic, physical component of white light.[26] However, the Newtonians did not deny that the eyes could play tricks. And as Johann Wolfgang von Goethe argued at the beginning of the nineteenth century, and the Hering school echoed a century later, Newton (and especially eighteenth-century Newtonians) overemphasized the physical nature of color at the expense of physiological effects.[27] Uncertainty about the sources of abnormal color vision thus never disappeared in the nineteenth century. The methodological difficulties that color perception posed to the work of natural philosophy help to clarify why color blindness became such an important issue.

Achromatizing the Body's and Telescope's Lenses

The development of the achromatic lens for scientific instruments provoked similar uncertainty about the nature of color. Before the mid-
eighteenth century, lenses refracted light in such a way that its constituent
colors were dispersed, giving microscopic and telescopic images an "un-
natural" look. In his public writings at least, Newton had been skeptical
that an achromatic lens for optical instruments could ever be developed.[28]
As Keith Hutchison has shown, this doubt arose largely from Newton's
neoclassical belief that matter was uniform in nature rather than diverse.
John Dollond's 1757 successful construction of an achromatic lens using two
different kinds of glass (English crown and German flint) challenged these
assumptions:

> In post-Dollondian optics, individual media seemed to exert idiosyncratic influ-
> ences on the rays that passed through them, and refraction forced a re-
> introduction of specific causes like the question-begging qualities of the once
> discredited Peripatetics. The central belief of the mechanical philosophy that
> matter is ultimately all the same was similarly cast into serious doubt: matter has
> to exist in a variety of individual forms, at least two of which react differently
> with light. Simple aetherial accounts of refraction thus became far less plausible,
> for light seemed to be reacting with the very matter of the lens, and since it re-
> acted differently with different matters, refraction was more a chemical interac-
> tion than a mechanical one.[29]

Natural philosophy demanded that the idiosyncrasy of natural phenomena not
be erased in the legitimate search for underlying, unifying forces. British philosophers tended to care less about preserving the uniqueness of phenomena,
but shared the recognition that in reality, vast numbers of them remained impervious to predictive law. Many sciences still could not make reliable predictions, and perhaps never could. In fact, the very study of achromatism had emphasized this point. Bavarian superiority in glass production from about the
1820s forced British artisans and experimenters to acknowledge the imperfection of their own local prisms and lenses.[30] Ironically, then, one of the most
important improvements to scientific instrumentation of the early industrial
period forced the recognition that optical readings would always have a localized, subjective element.

Idiosyncratic tools and results were seen as unfortunate, but all-too-real features of most areas of natural philosophy. Despite the confident air of most of John Herschel's *Preliminary Discourse on the Study of Natural Philosophy*, for example, he had to admit that vibratory and undulatory phenomena such as sound and light continued "to give . . . occasion for fresh researches; while phenomena are constantly presenting themselves, which show how far we are from being able to deduce all the particulars, even of cases comparatively simple, by any direct reasoning from first principles."[31] William Whewell went farther and stressed the differences between, rather than the unity of, the sciences. Like Herschel, though, he considered the least deductive sciences to be the least mature and progressed.[32]

Work on achromatism in the early nineteenth century, in sum, emphasized the growing complexity of optics. Neither light nor vision seemed explicable under simple, mechanical laws. Light's varying behavior through different media suggested the possibility of chemical reaction and it was generally unclear until several decades into the nineteenth century how light interacted with objects to make those objects seemed colored.[33] With its own unique set of lenses, the eye transformed these reactions into nervous (possibly electrical) impulses. Given the unique optics of human vision and the importance of the mind in the final stages of perception, natural philosophers suspected that physiological optics might require distinct laws not covered by physical optics. Individual visual discrepancies such as those discussed in the previous section reinforced this sense that human vision defied (and perhaps always would defy) any attempts to subsume it under universal law. This had an important implication: visual phenomena were still quite local.

This threat of enduring provincialism plagued natural philosophy, and goes a long way toward helping us understand the fascination with color blindness in the nineteenth century. The desire to comprehend color blindness stemmed not only from a broader scientific impulse to find the universal laws underlying natural complexity. It also arose from the crucial place of color vision in scientific methodology. Admitting that color vision was fundamentally idiosyncratic had the potential to frustrate one of the most important goals of British natural philosophy: the drive to subsume provincial understandings under national, unified systems. From its beginnings in the late eighteenth century, then, the scientific study of color blindness was a project in subduing that threat.

The Scientific Study of Color Blindness

An explanation of this curious defect will be worth while listening to, the more so as many eminent philosophers have suffered from it; and it is perhaps owing to this circumstance that so much time and attention has been given to the investigation of so curious an anomaly.

—Jabez Hogg, "Colour-Blindness"

Causes

John Dalton and his brother shared their color blindness, just as most other subjects shared the condition with a family member. From the beginning, then, color blindness was recognized as congenital.[34] Beyond its hereditary nature, early investigators knew very little about the cause(s) of color blindness, and the speculative nature of their theories plagued a natural philosophical community that prided itself on its avoidance of metaphysics.

Theories about the causes of color blindness tended to locate the problem either in the eye itself or in the rest of the nervous structure that supported vision (e.g., the brain, optic nerve). As I will discuss later, Dalton suggested that his vitreous humor was blue, and thus located the cause of his color blindness in his eye, but this idea was quickly disproved after his death. Brewster attempted to keep alive theories that located the cause of color blindness in the eye, not because he found them entirely convincing, but because no evidence had yet proved their falsehood.[35]

But most sought the cause of color blindness in the broader nervous and mental network of perception. Based on George Combe's study of a color-blind brass founder in Edinburgh, phrenologists argued that part of the brain in the middle of the superciliary ridge contained the faculty of distinguishing colors.[36] Those natural philosophers and physicians skeptical of phrenology pointed to the lack of correlation between such a structure and the occurrence of color blindness, but were at a loss for positive evidence for their own cerebro-retinal theories.

The early industrial age's most popular explanation of color blindness came from Thomas Young, a physician who became famous for his experimental support for the wave theory of light. Young believed that the normal eye had three vibrating systems, which responded respectively to red, green, and blue. In the color-blind, he argued, one of the three receptors did not function. Young's theory of color vision depended on the newer sense of perception as an active process. The retina, he argued, could only respond to stimuli in a finite number of ways, regardless of the infinite number of colors in the spectrum. The eye, therefore, acted as an analyzer or synthesizer of visual information, rather than a passive camera obscura.[37]

Young's hypothesis that color blindness might arise from an absence or paralysis of one of three color receptors in the eye lacked anatomical confirmation. At midcentury, Brewster grumbled the "humiliating confession that we do not even know that the retina is the seat of vision, and that, if it is, we know little of its structure or its functions."[38] He even conceded that apart from Dalton's disproved theory, all others were "mere conjectures" *incapable* of proof or disproof. Partly because of the lack of reliable stains and mounts, knowledge of the anatomy and physiology of vision proceeded at a pace frustrating to many opticists.[39]

Another popular position in the mid-nineteenth century—held most prominently by John Herschel and Elie Wartmann—attributed color blindness to a "defect in the sensorium." Anatomical theories such as Dalton's, they argued, did not account for the fact that a color-blind retina did register all colors as light. Such a retina simply did not distinguish certain refrangibilities as separate colors. The defect behind color blindness lay in the sensorium, "by which it is rendered incapable of appreciating exactly those differences between rays on which their colour depends."[40] Since the late eighteenth century the *sensorium commune* had replaced the soul as a convenient explanation for those dualists who could not quite bring themselves to attribute all mental activity to matter.[41] Except for a few die-hard materialists such as Joseph Priestley, most agreed that the mind ought certainly to maintain a status independent of the body. But arguing that color blindness was a sensorium defect could also be seen as merely positing the circularity that "the color blind have the defect of not seeing colours."[42]

Whether they believed the cause of color blindness resided in the eye or the mind, natural philosophers thought that scientific protocols would provide the discipline to both comprehend and manage this subjectivity.

Experimental Methods

Natural philosophers employed an array of techniques to test color vision. The earliest studies required subjects to describe or match the colors of pieces of cloth or glass. Dalton tried using the spectrum created by white light refracted through a prism, but found ribbons more convenient for transporting to other subjects and his public lectures. The search for ever-more precise methods of producing and naming colors led also to polarized light. Since many of those interested in vision in the eighteenth and nineteenth centuries also studied optics, they knew that polarized light produced especially pure colors. In the 1810s-'50s, mass-produced color charts such as those of Abraham Gottlob Werner, David Ramsay Hay, and M. E. Chevreul appeared.[43] These charts allowed investigators to compare their subjective experience to standard measures.[44] Finally, by midcentury, instruments such as Maxwell's color box, and his mathematical techniques for expressing color vision, married transportability with precision measurement.[45]

One can see a similar change during the early industrial age in the operative definitions of observer and subject. In much early modern physiological and psychological research, observer and subject were the same person. In some important cases, the main investigator did not trust his vision enough to stand as the primary observer. Newton, for example, admitted as much in his *Opticks*. To correlate the musical octave to the visible spectrum, he needed to know where on the spectrum one color stopped and another began. He had "an Assistant, whose Eyes for distinguishing Colors were more critical than mine" draw these lines.[46]

By the same logic, the authority of the unique case history, which had proven so popular in Enlightenment natural philosophy, in the early nineteenth century became subverted to the search for causes, statistical prevalence, and other more generalized accounts of abnormality. As I will explain, the attempt to understand color blindness and the attempt to absorb provincial knowledge into natural philosophy adopted the same strategy. By midcentury, investigators of color vision had at their disposal a standard set of equations, experimental techniques, and national statistics that allowed them to conceive of their research as a national, even international concern.[47] Thus, during the early industrial period natural philosophers mounted a methodological case that they could successfully manage the subjectivity of color blindness — and by extension that they could manage other issues as well.

Earliest Reported Cases

Several scattered cases of color blindness had been reported before Dalton's 1794 paper.[48] The first extended, published account of color blindness in 1684 had described a twenty-three-year-old woman who saw only in black and white.[49] Soon after, French engineer-mathematician Philippe de La Hire and English chemist Robert Boyle each mentioned the condition in passing in their optical works.[50]

Only in the late eighteenth century, however, did several published papers address color blindness as an issue in its own right. In 1777, Joseph Priestley read to the Royal Society of London a letter from Joseph Huddart concerning "Persons who could not distinguish Colours." Huddart described an acquaintance, a Quaker shoemaker named Harris, whose untimely death prevented Huddart from conducting his own tests. The case history as Huddart told it had features that many subsequent accounts would share: Harris had noticed a difference between his and others' color vocabulary; he first suspected such a difference when a stocking he found as a four-year-old was described by others as "red"; he could not distinguish cherries from leaves on a tree; his vision otherwise seemed normal; two brothers shared his defect. Huddart was able to test one of Harris's brothers, who worked on a trading vessel. Not having a prism at hand, Huddart asked the man to name the colors on a variety of ribbons. He mislabeled orange and red the most seriously.[51]

Some months later, another instance of color blindness trickled up to the president of the Royal Society, John Pringle. The Huddart paper had inspired Stephen Whisson, the university librarian at Cambridge, to follow up on a case of color blindness of which he knew, one J. Scott of Lincolnshire. Like Harris, Scott could reliably identify only the colors yellow and (in most cases) blue. His discovery of his condition also involved clothing: he was mistakenly offended when the man who was to marry his daughter arrived one day wearing what appeared to be an inappropriately somber black suit. The young gentleman actually wore the claret color then common to those in the legal profession. Unusually (as it would turn out), Scott shared his color blindness with not only several males in his family but also some female relatives.[52]

Scottish Common Sense philosopher Dugald Stewart's red-green color blindness also piqued contemporary curiosity. One could have learned from the *Edinburgh Encyclopaedia* that neither Stewart nor those close to him noticed his color blindness until it "was discovered from the accidental circumstance of one of his family directing his attention to the beauty of the fruit of the Siberian crab, when he found himself unable to distinguish the scarlet fruit from the green leaves of the tree." University of Edinburgh medical professor William Pulteney Alison, son of a well-known author on taste, remarked that he had "seen Mr. Stewart look at a scarlet geranium in full flower, and say that he was not sensible of difference of the colours of leaf and flower."[53]

Stewart may well have felt embarrassed by his color blindness, for he never mentioned it publicly and at one point drew on the trope that blindness meant provincialism. He considered the cultivation of *all* aspects of philosophy and the arts as important to the individual's moral development. Specialization was anathema to the ultimate pursuit of moral philosophy. He posed this analogy to drive his point home: "A man who loses his sight improves the sensibility of his touch: but who would consent, for such recompense, to part with the pleasures which he receives from his eyes?"[54]

These early accounts of color blindness hinted at several themes that in the early nineteenth century would become full-blown issues: the embarrassment to a natural philosopher of having subjective vision and the social errors that color blindness could cause. Not until Dalton brought color blindness to national attention, however, would the visual abnormality become an opportunity to publicly test the management skills of natural philosophy.

The Paradigmatic Case:
John Dalton and Provincial Genius

John Dalton first noticed his unusual color perception in his garden, which he had been cultivating since 1790 for botanical research. More than his previous work, this research required close scrutiny of color, and he began to notice that he found certain color distinctions difficult to make. "With respect to colours that were white, yellow, or green," he said, "I readily assented to the appropriate term; blue, purple, pink, and crimson, appeared rather less distinguishable, being, according to my idea, all referable to blue. I have often seriously asked a person whether a flower was blue or pink, but was generally considered to be in jest." Still, Dalton did not seriously suspect that his vision was peculiar until two years later, when he noticed that a pink geranium appeared blue in day-light but red by candle-light. Only when others could not see this curious phenomenon did Dalton realize that his eyesight "was not like that of other persons."[55]

His career had just begun to take a turn for the better. From the beginning, Dalton's choices had been limited by the accidents of cultural and geographic place. Born into a family of modest means, he had been teaching since the age of twelve. When Dalton and his older brother Jonathan opened a co-educational school together in 1785, it thrived reasonably well, but seems to have lost students due to the Dalton brothers' "uncouth manners" resulting from their lack of "intercourse with society."[56] At twenty-four, the younger Dalton was piecing together tutoring jobs in Kendal, despairing all the while of forging a comfortable life in natural philosophy. An uncle in London discouraged the young Dalton's hopes of becoming a physician or barrister with the blunt observation that Quakers usually found those professions "totally out of [their] reach."[57]

He hoped to find more stability in Manchester, where he moved in 1793. On the recommendation of his mentor, John Gough, a Unitarian establishment there called the New College had invited Dalton to work as a tutor in mathematics and natural philosophy. After six years at the college he retired to a living as a private tutor and researcher in a laboratory that the local literary and philosophical society allowed him to keep in one of their rooms.

Despite his fairly comfortable circumstances and circulation in "good" society, Dalton's biographers would find great dramatic material in this tension between genius and provincial circumstance. Fellow Cumberland native Henry Lonsdale, for instance, marveled that such a dramatic change in chemical thought as the atomic theory could come from such "a very unexpected quarter of England—a city of cotton interests and hard cash, not without laudable ambition to become 'the Cottonopolis of the North.' The lamp of knowledge got trimmed amid the din of shuttles and spinning-jennies and multifarious handicrafts by an unobtrusive Quaker, pursuing his calling of schoolmaster in a back street of Manchester, and thankful to earn the wages of a skilled artisan. Yet this humble individual, scarcely known outside the pale of his peculiar religious denomination, was daily absorbed in profound intellectual studies."[58]

Dalton's poverty at that point in his career is highly dubious,[59] and it seems far more likely that he was lionized for representing the transcendence of genius over provincial limitations rather than pennilessness. But Dalton himself, in some accordance with Lonsdale's portrayal, did relish his provincial status. For instance, he declined a nomination to the Royal Society and reserved the announcement of his atomic theory for Edinburgh and Glasgow rather than London.[60] Commenting on the latter decision, he said he figured that "the doctrines inculcated would in those cities [Edinburgh and Glasgow] meet with the most rigid scrutiny, which is what I desire."[61] He also insisted to Roderick Murchison at the first meeting of the British Association that the members should always remain "*Provincials,*" lest they "lose all the object of diffusing knowledge."[62]

Likewise, when Dalton first entered the public world of natural philosophy with his 1794 color blindness paper at the Manchester Literary and Philosophical Society, he sought to take advantage of a newly vibrant provincial market.[63] He told his audience that he had first tested his sight on the solar spectrum, and found that he only distinguished two colors (yellow and blue) where others saw six or seven. "My yellow comprehends the *red, orange, yellow,* and *green* of others," he said; "and my *blue* and *purple* coincide with theirs." Dalton had soon found that his brother and twenty others in Eaglesfield and Manchester, including two of his pupils, reported similar experiences.[64]

The cause of this condition perplexed Dalton, until he noticed that candle-light passing through a "sky-blue transparent liquid" looked like daylight. He boldly asserted "almost beyond a doubt" that the color-blind must have an extra, blue-tinted vitreous humor in the eye which altered the red, orange, and green parts of the spectrum.[65] To test this hypothesis, he asked his medical attendant Joseph Ransome, to dissect his eye after his death (see Figure 3.1). Ransome found, contrary to Dalton's hypothesis, that the vitreous humor was colorless and that the crystalline lens had merely taken on the amber hue typical of old age. The attendant then cut open the other eye, and viewed light through it as if through a lens. He noticed no *"appreciable difference."* When David Brewster visited shortly thereafter, Ransome encouraged him to perform the same experiments, and they agreed "that the imperfection of Dalton's vision arose from some deficient sensorial or perceptive power, rather than from any peculiarity in the eye itself." Phrenology also tested its claims on Dalton's cadaver. One of Johann Gaspar Spurzheim's former assistants, Bally, got permission to take a cast of Dalton's head during Ransome's investigation, and claimed to have found a deficient "organ of color" at the frontal sinus. A decade later, George Wilson argued that Dalton's large brow discredited the deficient organ theory.[66]

Quakerism and Color Blindness

Nineteenth- and twentieth-century writers have enjoyed relating again and again the story of Dalton's color blindness. More than the dispute about color blindness' cause, the fact that Dalton was a Quaker has seized the imagination. Indeed, the Society of Friends' place in British culture had an important part to play in the understanding of color blindness. Not only was the most prominent case a Friend, but so were a number of other cases reported during the early industrial age, and one of the major theorists on the cause of color blindness, Thomas Young.[67]

It is not immediately obvious why religious affiliations should have mattered to the natural philosophers and physicians who studied this condition. I contend that Quakerism struck natural philosophers as relevant to studies of color blindness because both qualities symbolized provincialism. The cultural strictures imposed by the Society of Friends seemed very strange to mainstream Anglican Britain, even to many other dissenters. Both the Quaker and the color-blind were understood to sit at a remove from (religious or visual) truth because they did not properly understand or communicate it. The color-blind person did not understand because he could not; the Quaker because he

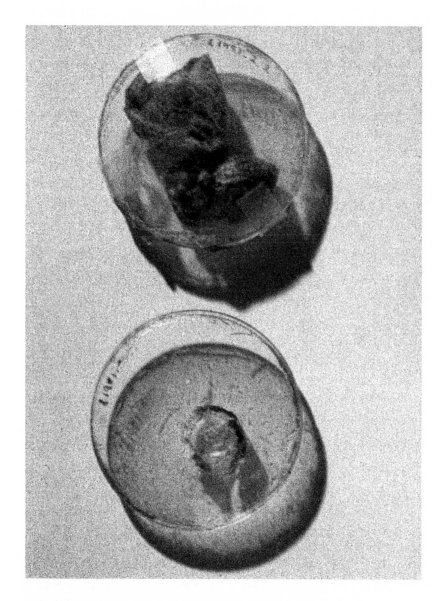

FIGURE 3.1
Ransome Tests Dalton's Theory
Remarkably, the Manchester Museum of Science and Industry still keeps in its collection John Dalton's eyeballs — an eerie relic both of Dalton's mistaken theory of his own color blindness and of the totemic quality of his person. Photograph courtesy of the Museum of Science and Industry in Manchester.

would not. Whenever a color-blind subject happened to be Quaker also, then, this fact seemed worth reporting. For example, Joseph Huddart, in his 1777 description of the color-blind shoemaker, had speculated that the difficulty Harris had in noticing his unusual eyesight stemmed in part "from the circumstance of his family being quakers [*sic*], among whom a general uniformity of colours is known to prevail."[68] This paper provoked Dalton's realization that his case was not unique, and he had a friend test the living members of the Harris family with a detailed set of questions and colored ribbons.[69]

Two aspects of traditional Quaker culture—their dress and their language—seized the British imagination and also bore the most relevance to color blindness. One color-blind subject, Frederic Dyster, told Brewster that while he would perceive a woman in light blue as gaily dressed, "habited in pink, she might pass as a Quakeress," since pink appeared as a "dirty slate-colour" to him.[70] When George Fox first formed the Society of Friends in the mid-seventeenth century, devout Protestants commonly wore muted colors, and the Quakers adopted and codified this practice. They considered drab clothing less idolatrous both because it called less attention to its wearer and because it cost less. Though there was no specific uniform, Quakers were expected to dress according to "decency and comfort," and "gay colours" such as red, yellow, and blue were prohibited. Dalton's own father, Joseph, had earned a living weaving the gray cloth that many Friends preferred.[71] As William Penn admonished in 1757, "Visible Objects have a great Influence on The People, and therefore Satan is represented tempting Eve with fair Fruit, pleasant to the Senses, the Palate, the Eye, etc. And are not they who sell vain and superfluous Things, exposing them at Door and Windows, Tempters of the People to the Lust of the Eye and pride of Life."[72] The Friends' belief that aesthetic cultivation through clothing, dance, and other expressive forms interfered with spiritual growth directly contradicted the conventional wisdom in mainstream British culture that the cultivation of taste made one a more well-rounded and even healthier person.[73]

The Friends' rules of language also reflected their desire to eschew heathenism. For instance, the colloquial use of "you" to refer to one or more persons (originally "you" was the plural of "thou") Fox found not only grammatically incorrect but also idolatrous because it implied that the individual spoken to was important enough to be addressed as more than one person. Likewise, Quakers refused to use even the most generic titles, and called each other "Friends" and outsiders "Neighbours."[74]

These customs attracted a great deal of scorn. For example, while Quakers defended their changes to the language as truer, more logical, and closer to divine teachings, to some critics this grammar seemed like a reversion to the Tower of Babel. The Tory poet laureate Robert Southey found their speech "vague and rambling" despite itself. A common rejoinder stipulated that, on the contrary, Quakers had eliminated much of the ambiguity and double meaning from the English language.[75]

Still, in the nineteenth century a number of Friends chose to relax the grip their religion had on their secular life. The early nineteenth century saw a growing factionalization between traditional and evangelical forms. Traditional, or "quietist," Friends emphasized turning inward in the hope of understanding the "Inward Light," the divine spirit in every person. The evangelical faction, which came to dominate the Society over the course of the century, emphasized the external conditions of Jesus's life and encouraged great works (e.g., involvement in the antislavery movement) rather than mysticism.[76]

This context is crucial to understanding Dalton's case and how it was interpreted in the nineteenth century. To his contemporaries, the Mancunian was enough of a Quaker to serve as an object lesson in subjectivity and provincialism but also relaxed enough in his religious practice to transcend otherwise narrow circumstances. The exact nature of Dalton's religious observance is unclear, but he did donate money to the Society, regularly attended meetings, and observed the plain dress of his fellow Quakers. He willingly loosened his stricter habits when with non-Friends, but he maintained a serious interest in simple language. This perhaps is not surprising, given the issue's importance for not only Quakerism but also post-Lavoisierian chemistry. Soon after he investigated his color blindness, Dalton published a successful book on English grammar that emphasized the importance of economy in language.[77]

That Dalton was not rigidly quietist did not preclude his contemporaries from calling attention to his faith. In fact, his religion provided a context both for the discovery and public understanding of his color blindness. Before he had fully realized the nature of his vision, one possibly apocryphal story went, Dalton bought what he thought were somber-colored silk stockings for his mother. When he showed them to her, she protested that she could never wear anything dyed such a bright red. Not believing he could make such a gross mistake, Dalton called in his brother, who denied equally vehemently that they were red. Deborah Dalton had to gather reports on the stockings' color from several women friends before her sons would agree that they themselves had the deficiency.[78] (Women, as we have seen, had an additional provincialism to overcome, namely the epistemological limitations of their own gender.)

Even more amusing to nineteenth-century commentators was Dalton's attire at the 1832 British Association meeting in Oxford. The university took the opportunity to award honorary D.C.L. degrees to the chemist and several other men of science. Dalton arrived wearing a scarlet gown. "As it appeared to him of the same modest colour as the foliage around him," wrote fellow degree recipient Brewster, "he was not aware of the brilliancy of his plumage, though he often jocularly referred to his incapacity of appreciating it." When he donned the same attire to receive an honorary LL.D. from the University of Edinburgh two years later, several Anglican bishops in attendance expressed horror at such a violation of the modesty and gravity that they expected from a philosopher—especially a Quaker one. William Whewell recalled asking Dalton to point to an object whose color matched his scarlet doctor's robe. The chemist pointed to the trees.[79]

In his last and, to his contemporaries, most amusing public performance, Dalton was presented to King William IV, thanks to the machinations of Charles Babbage. Quite sensitive to the delicacy of the matter, Babbage found Dalton not only willing to endure a ritual that might offend a more traditional Friend, but also amenable to wearing again the scarlet robes of an Oxford Doctor of Laws (since the traditional court dress and sword were even more inappropriate for a Quaker). After much coaching from Babbage, Dalton appeared at court, and again attracted significant attention because of his dress and religion. When the latter became apparent to the party at court—which included the Archbishop of Dublin—Babbage's reassurances barely assuaged the general feeling that "the Church [was] really in danger."[80]

The comedy of these stories lay just as much in the hilarity of a humble, Manchester Quaker appearing at court, as in his mistaking of colors. Mixed with the humor, though, was the pride that Brewster, Whewell, and Babbage felt that a scientific man could break through traditional social boundaries. Indeed, each of these stories was meant to convey the idea that natural philosophers required their own rules of discipline, unshackled from the restraints of illiberality. Each Dalton anecdote ended with the same serious moral: this man managed to steer a moderate path between the Scylla of the sterner strictures of quietist Quakerism, and the Charybdis of conservative Anglicanism. He adopted whatever productive qualities that Quakerism offered ("stern self-reliance," "indomitable perseverance," "characteristic simplicity") but moderated them. Thus, he did not allow his "usual dryness and quaintness of style" to stifle his "imagination and the power of painting with the pen." If his thoughts and actions rarely ran explicitly to national concerns, they were all the more clearly directed at the abstruse pursuit of chemical theory.[81] The humility of Dalton's dress and speech reflected his pure devotion to scientific work without regard for rewards or honors.[82]

In composing his polemic, *Reflections on the Decline of Science in England*, Babbage realized the rhetorical power that Dalton's story promised. Here was a man, so crucial to the revival of chemistry in Britain—who had to continue teaching well into his seventies just so he could live modestly. It was an outrage. With the help of Dalton's friend, William Henry, as well as the publisher J. A. Murray and two Manchester members of Parliament, G. W. Wood and C. Poulett Thomson, Babbage campaigned for a pension for Dalton.[83] Finally, at the 1833 British Association meeting in Cambridge, Adam Sedgwick announced that Lord Grey's government had conferred on Dalton a modest £150 pension, which three years later increased to £300. In a long testimonial that was submitted to the government, Henry portrayed the chemist as a new kind of heroic scientific figure, one who disciplined his search for knowledge with modesty:

> By the moderation of his wants, and the habitual control over his desires, he has been preserved from worldly disappointments; and by the calmness of his temper and the enlargement of his views, he has escaped the irritations that too often beset men who are over anxious for the possession of fame, and are impatient to grasp prematurely the benefits of its award....
>
> He has . . . secured our admiration not for *ingenious opinions* merely, but for having worked out the evidences of these opinions by labour most sagaciously and perseveringly applied.[84]

Dalton's isolation had protected him from avarice and, ironically, had resulted in the "enlargement of his views." Himself a Mancunian, Henry therefore provided a powerful counterpoint to the provincial-as-myopic model. For Henry, the point was to honor and manage provincial knowledge, not to deride or stifle it. The very fact of Dalton's provincialism combined with his scientific acumen made him more, not less, capable of attaining truth.

Henry's depiction of Dalton as a scientific diamond in the provincial rough met with some resistance, though. The fashionable London physician Henry Holland found Dalton's 1807 popular lectures in Edinburgh exceptionally unsophisticated: "His experiments, as made in public, frequently failed. His voice was harsh, indistinct, and unemphatical; and he was singularly wanting in the language and power of illusion, needful to a lecturer in those high matters of philosophy, and by which Davy and Faraday have given such lustre to their great discoveries."[85] Humphry Davy himself found the Manchester chemist to be "a very singular man. A Quaker by profession and practice; he had none of the manners or ways of the world." As the president of the Royal Society presenting the first-ever Royal Society medals to Dalton and James Ivory, and as an ambivalent figure in the Royal Society reform attempts in the 1830s, Davy depicted his fellow chemist as a provincial whose limited knowledge of the latest research owed much to his more sophisticated metropolitan counterparts, Thomas Thomson and William Hyde Wollaston.[86]

Similarly, George Wilson, who would write the century's most comprehensive text on color blindness, claimed that the Friends showed "a greater proneness to colour-blindness than any other class or denomination," and even suggested that their color receptors had, from centuries of disuse, lost their sharpness in Lamarckian fashion. Wilson predicted that color blindness might disappear by the end of the century along with the Quakers, whose numbers were already diminishing.[87] An American reviewer sympathetic with the British reformist position found this idea preposterous. "Unfortunately for this theory Quakers are not always looking at their clothes, nor are they shut out from the varied hues of nature and art, nor does their defect bear any distinct relationship, complimentary or otherwise, to the prevalent drab of their denomination. The fact that Dalton was a member of their persuasion, and that consequently minuter researches may have been instituted amongst the body, will explain why they have furnished so large a contingent of patients."[88] The writer also suggested that Dissenters' very heterodoxy might better prepare them to notice a visual defect. He imagined trying to convince an assembly of a thousand people that in all probability twenty of them were color-blind. Their alarm at this idea, he said, could only be answered with the uncomfortable notion that the human eye was vulnerable to all kinds of "chromatic heresies, although the owner might think himself as orthodox in vision as every man deems himself in the Faith."[89] The purported provincialism of dissent here met with a powerful counterargument shared by many scientific reformers: if properly framed by controlled scientific study, those outside the mainstream could make substantial contributions to natural knowledge and society at large.

In sum, John Dalton presented a fascinating and complicated case to a natural philosophical community that was highly preoccupied with the expansion of its base. Those threatened by provincial science tended to depict Dalton as just a few steps above a country bumpkin. His color blindness and his Quakerism symbolized the limitations on provincial knowledge. For scientific reformers such as Brewster, Henry, and Babbage, however, these features served only to dramatize Dalton's transcendence. Color blindness, humble living as a tutor, the simple speech and dress of a Friend—none could stifle (and some might even encourage) that rare breed of native genius, skill, and perseverance that could arise in any part of the nation. This was an idealized picture, of course. In reality, Dalton was not quite so humble, nor were local knowers quite so trusted, even by reformist natural philosophers. In the early nineteenth century at least, the discussions of Dalton's character and color blindness did more to illustrate the desires than the actual achievements of a scientific reform movement.

STANDARDIZATION AS A SOLUTION FOR PROVINCIALISM

In the mid-nineteenth century, scientific reformers turned wishful thinking into considerable success by developing standardized methods for studying color blindness. Before about 1830 the literature on color blindness had assumed the form of individual case histories such as Harris's or Dalton's, rather like those found in the medical or antiquarian literature. Indeed much of the early interest in the problem focused on it as a curiosity rather than an anatomical, physiological, or psychological issue ready for extensive experimental study.[90] As more and more cases poured into philosophical, medical, and popular journals, however, it became clear to natural philosophers that they needed to turn their attention to finding causes and developing more standardized procedures.[91] Particularly as popular interest in the sciences increased, natural philosophers found it necessary to differentiate knowledge of mere individual facts from the more prestigious pursuit of finding the "laws and generalizations by which they are connected, and the appropriate ideas or theories by the assistance of which a body of science is constructed and united."[92] Post-Daltonian investigators of color blindness sought ways to standardize their tests and language. They aimed to discover the laws and perhaps even the causes behind this visual anomaly, and in the process learn something about normal vision as well. In other words, in the early nineteenth century, those who studied color blindness sought to transcend the first stage of induction—gathering subjective cases—to develop quantitative, causal, universal laws of human perception. They could thus put the study of vision on the same footing that, for instance, astronomy had already achieved through celestial mechanics. The science could transcend its provincial status.

Lorraine Daston has argued that the mid-nineteenth century witnessed the birth of "aperspectival objectivity," the attempt to eliminate individual idiosyncrasies from scientific work. Even before Gustav Fechner and Wilhelm Wundt attempted fully to mechanize the psychological subject in the 1870s, frustrations with the idiosyncrasy of individual experience provided strong motives to find more "objective" testing techniques. The study of psychology and nervous physiology might never rid itself entirely of subjective introspection, the thinking went, but it could discipline it with standardized procedures and instruments.[93] Another impetus for the development of aperspectival objectivity was the much earlier "disappearance of the patient's narrative" in medicine. The growing independence of the medical profession in the late Hanoverian period allowed it to rely less on the patient's possibly idiosyncratic and unreliable narrative of their illness, and concentrate instead on signs and symptoms.[94]

These new values of aperspectival objectivity and a more standardized subject required that knowledge translate across local boundaries and levels of training and skill. The kind of unique experience that Dalton reported in his color blindness paper seemed quaint to those mid-nineteenth-century men of science who operated under the "ethos of the interchangeable and therefore featureless observer."[95] Statistical and other quantitative methods entered vision studies more slowly than other sciences. When they did infiltrate vision studies, however, abnormal conditions such as color blindness became quantifiable, imminently controllable variations of normal vision, rather than more curiosities for the cabinet.[96]

In what follows, I will demonstrate a shift from anecdotal reporting of color blindness cases to an understanding of color blindness that drew from the conventions of aperspectival objectivity. At midcentury, engineer George Wilson conducted the first statistical survey of the defect's prevalence in Britain, and simultaneously James Clerk Maxwell and others developed more systematic methods for testing color vision. The statistical survey redefined color blindness as a nationwide phenomenon, while the new testing procedures seemed to eliminate idiosyncratic testing or reporting. With the help of those with "normal" eyesight, the color-blind were left with an "objective" rendering of their visual abnormality.

An early signal of this change from case history to standardized study came in 1824. That year, George Harvey presented a case of color blindness to the Royal Society of Edinburgh. His reporting of a single case was hardly unusual. But his testing methods harkened something new. A chart from one of

Patrick Syme's popular editions of Werner's *Nomenclature of Colours* offered him a standardized instrument that could be employed by any subsequent observers.[97] In this respect, the technique marked a step away from the discovery narratives of Dalton and his successors, and toward a universal standard by which color vision could be judged.

Another Provincial Genius: Edward Troughton

John Herschel took the standardization of color blindness studies a step farther in his account of Edward Troughton. The study appeared in Herschel's treatise, "Light," the Ur-text of British mathematical wave theory. As one of the nation's finest precision instrument makers, Troughton represented the new world of precision at the service of nonprovincial, mathematical theory. One can hardly exaggerate the extent to which Troughton's reputation relied on unusually acute vision combined with extraordinary manual skill. Troughton's acclaimed method for manually dividing transit circles that were too large for a dividing engine required him scrupulously to make marks and corrections using his eye, hand, microscope, and a tapered roller marked into segments. It is thus remarkable to find such a success story in a man who not only had limited color vision, but also had lost an eye in an accident.[98]

Troughton and his business partner, William Simms, were mathematical rather than optical instrument makers, so it was not unusual for them to commission any lenses that they needed made to others, namely George Dollond and Charles Tulley.[99] Still, Troughton's color blindness does seem to have had one important effect on his work. At least one contemporary, Richard Sheepshanks, remarked that "the only astronomical instrument which is not greatly indebted to Mr. Troughton is the *telescope;* and he was deterred from any attempt in this branch of his art [by his color blindness]. . . . With this defect in his vision, he never attempted any experiments in which colour was concerned; and it is difficult to see how he could have done so with success." But Sheepshanks quickly pointed out—as did most biographers of visually impaired natural philosophers—that Troughton's color blindness did not indicate otherwise defective eyesight. No one could find lacking in Troughton the acute visual skills required to construct some of the finest optical instruments in the world.[100]

A fellow of the Royal Society of London, Troughton associated socially and professionally with the most prominent natural philosophers of his day. At the Royal Society, he presented papers whose mixed mathematical character helped him transcend the role of "mere" instrument maker. The Cumberland native's refined skill and learning interested his contemporaries all the more because they were "better than usual for his rank." While George Airy privately mocked the "black dirty old man in his dirty wig and black dirty shorts," he admired Troughton's genteel liberality in discussing his techniques with outsiders. In short, Troughton was yet another example of a classic provincial figure gone cosmopolitan.[101]

In its outlines, the instrument maker's rise from provincial obscurity to cosmopolitan fame paralleled Dalton's. So when John Herschel made Troughton's color blindness widely known, the case became a logical successor to Dalton's. Following the anonymity conventions typical to case studies at the time, Herschel did not name his subject. But his identity was no secret to the scores of natural philosophers who read the paper.[102]

For his tests on Troughton—probably performed in 1823[103]—Herschel developed a new, more precise protocol in which he polarized light through an inclined plate of mica to produce two circles. He could then vary the color of the two circles by changing the inclination of the mica, but at any one time the two circles were always of complementary colors. He asked Troughton to describe each color in his own words, and compared the responses to his own perceptions. He also had Troughton adjust the mica himself to produce perfectly contrasting colors (see Figure 3.2). Both trials made it clear that Troughton could only "fully appreciat[e] blue and yellow tints, and that these names uniformly correspond, in his nomenclature, to the more and less refrangible rays, generally; all which belong to the former, indifferently, exciting a sense of 'blueness,' and to the latter of 'yellowness.'"[104] Even more importantly, Herschel invented an experimental technique that simultaneously provided a more standardized, transportable way to measure color blindness and limited such methods to those with sufficient training to understand polarization.

Through these inquiries, Herschel realized that Young's theory about the nature of color blindness—if not his postulated mechanism—made the most sense. While those with normal vision registered color as combinations of three primary sensations, those with Dalton's type of color blindness only had two kinds of sensation excited by light. Accordingly, Herschel labeled this defect "dichromic vision," but admitted that other types of color blindness probably existed. Unlike Young, Herschel preferred not to speculate where the seat of this dichromic defect lay, or which three hues specifically corresponded to putative receptors in the eye.[105]

Colours according to the judgment of an ordinary eye.		Colours as named by the individual in question.		Inclination of the plate of mica to eye.
Circle to the left.	Circle to the right.	Circle to the left.	Circle to the right.	
Pale green.	Pale pink.	Both alike, no more colour sky out of window.	in them than in the cloudy	89.5°
Dirty white.	Ditto, both alike.	Both darker than before, but	no colour.	85.0
Fine bright pink.	Fine green, a little verging on bluish.	Very pale tinge of blue.	Very pale tinge of blue.	81.1
White.	White.	Yellow.	Blue.	76.3
The limit of Rich grass green.	pink and red. Rich crimson.	Both more coloured Yellow.	than before. Blue.	74.9
Dull greenish blue.	Pale brick red.	Better, but neither Blue.	full colours. Yellow.	72.8
Purple (rather pale.)	Pale yellow.	Neither so rich Blue.	colours as the last. Yellow.	71.7
Fine pink.	Fine green.	Coming up to good colours, than a gilt picture-frame. Yellow, but has got a good deal of blue in it.	the yellow a better colour Blue. but has a good deal of yellow in it.	69.7
Fine yellow.	Purple.	Good yellow.	Good blue.	68.2
Yellowish green.	Fine crimson.	Better colours than Yellow, but has a good deal of blue.	any yet seen. Blue, but has a good deal of yellow.	67.0
Good blue, verging to indigo.	Yellow, verging to orange.	Blue.	Yellow.	65.5
Red, or very ruddy pink.	Very pale greenish blue, almost white.	Both gay colours, particularly Yellow.	the yellow to the right. Blue.	63.8
Rich yellow.	Full blue.	Fine bright yellow.	Pretty good blue.	62.7
White.	Fiery orange.	Has very little colour.	Yellow, but a different yellow, it is a blood-looking yellow.	61.2
Dark purple.	White.	A dim blue, wants light.	White, with a dash of yellow and blue.	59.5
Dull orange red.	White.	Yellow	White, with blue and yellow in it.	59.0
White.	Dull dirty olive.	White.	Dark.	57.1
Very dark purple.	White.	Dark.	White.	55.0

FIGURE 3.2
Herschel's versus Troughton's Color Decriptions
Using himself as an "ordinary eye," Herschel juxtaposed his own descriptions of colored light and Troughton's. Source of illustration: Herschel, "Light."

Toward Qualitative Precision: Herschel's Continued Studies of Color Blindness

During the 1820s and 1830s, at the time he was conducting his Troughton tests, Herschel was devoting most of his time to astronomy, optics, and photography. As with his father, William, John's studies occasionally led to questions about vision.[106] He offered the term *photology* to encapsulate these fields' common study of light and vision.

So Herschel's interest was piqued when, soon after publishing his much-admired "Light" essay in 1845, one Charles May wrote to say that he knew several color-blind people. May was one half of the instrument-making team Ransome and May, who constructed several of the devices used at the Royal Greenwich Observatory. Herschel thus knew he could trust May with a long set of instructions for experiments to perform with these subjects. Most of these experiments involved having subjects look at different sources of light through colored glasses and prisms that Herschel provided, and recording the colors seen. Some of the experiments required the aid of someone with both normal vision and some optical know-how, but Herschel asked for the subject's "habitual nomenclature of colours" so that the results would not be "biassed by the judgement of any bystander."[107]

Herschel's request that the subjects look at different kinds of light (direct and indirect sunlight, candlelight, light from oil lamp whose wick has been soaked in salt, a solar spectrum) drew from two potential sources. First, Dalton had perceived colors differently in different lighting conditions, and Herschel likely wished to know the prevalence and extent of this situation among other color-blind people. Second, it is also possible that Herschel suspected spectroscopy could help him understand color blindness. Natural philosophers with normal vision had analyzed the color "contents" of different light sources. A mapping of physical spectra onto abnormal physiological spectra might help pinpoint a chemical basis for color blindness.

At the very least, by comparing the experience(s) of the color-blind to the experimental readings of someone with normal color vision (presumably Herschel and May used themselves as controls), he could stabilize disparate case histories into one cohesive, objective understanding of color blindness.[108] That this was his intent was made clear in several ways. He took care to send the same set of questions, colored glasses, and other testing materials to each of his subjects. Also, Herschel asked Dalton to re-test himself rather than rely upon his older reports. Dalton's case would become much easier to compare with others if he answered Herschel's questions using Herschel's testing equipment.

Whereas Dalton handled his own tests, May knew his other subjects required expert assistance. For example, he had J. J. Lister—who had developed an achromatic lens for the microscope—conduct the tests on Joseph Foster, a seventy-year-old, color-blind man in Middlesex. To Dalton (and presumably to the others as well), Herschel also sent a large number of silk skeins, each of a different color, several colored glasses, and a prism. May himself tested a young man in his own neighborhood, though these notes no longer exist.[109]

The reports from these subjects were—perhaps needless to say—quite different. Whereas Dalton filled four pages with his own experimental notes, Lister noted that in Foster's case the "Questions were but partially + hastily answered for want of time." Foster skipped three of the six rather lengthy trials. More importantly, Dalton and Foster differed in their choice of color terms, despite the fact that they seemed to have had similar cases of red-green blindness. For instance, compare their descriptions of a refracted beam of daylight seen through two colored glasses:

Dalton	*Foster*
yields a band of Light, no blue in it	Like fog—a yellowish dirty white—... no blue
orange red band, no blue	not so white as the last—a little more dirty yellow—not so light[110]

These differences would have surprised neither Dalton nor Herschel, though this made them no less confounding. In the nearly forty years since he had conducted his original work on color blindness, Dalton wrote in his report, he had encountered a surprising number of other individuals (many of them his students) who "see nearly or exactly as I do." But having presumed a broad similarity, Dalton had to concede a nearly unavoidable subjective element to the experience of color blindness, a subjectivity that could only be contained through the tireless comparison and standardization of those experiences. He said of a color-blind acquaintance: "We agree both in words + ideas in all the grand points of distinction; but in some of the minor points we differed a little in *words*, though our ideas appeared the same upon mutual explanation."[111] Eventually, however, Dalton and his acquaintance learned to negotiate a common vocabulary for what they understood to be a shared abnormality.

This creation of a common vocabulary indicated the possibility of a more standard color language and therefore of a universalizable understanding of color vision. Herschel's cohort of reformist natural philosophers particularly hoped that standardized techniques might overcome the provincializing effect of language differences generally just as Dalton had done in his own conversations with his color-blind friend. As I discuss later in this chapter, Wilson and Maxwell would later supersede Dalton's and Herschel's qualitative precision by developing quantitative systems of language for vision studies. But negotiated vocabularies such as Dalton's were seen as a step in the right direction.

Herschel also gave a copy of his instructions and a duplicate set of colored glasses to William Whewell, who then took advantage of his acquaintance with T. R. Malthus's color-blind wife, Harriet. Her responses to Herschel's queries equally illuminate how problematic language could be in revealing the nature of color blindness. When asked how many colors she could distinguish, Malthus listed "1 Black, 2 White, 3d Yellow or Orange, 4 sky-blue, pink, 5 deep red, scarlet, bright green, such as a laurel leaf, bright brown, 6 dust of road, pea green, pony colour, 7 deep Blue, purple, violet, crimson, 8 blue or pink mixed with red, brown or green."[112] Her more precise appellations such as "dust of road" and "pony colour" instead of brown revealed both the level of detail achievable by subjects of color vision case histories, and also the qualitative nature of that precision. In the years surrounding Herschel's color blindness studies, other experimenters would seek to develop more intricate color wheels and charts to hone that precision, even if they did not immediately render that precision quantitative. At the time, Whewell regretted not having access to such tools, "for then we might have known to what part of the spectrum her descriptions referred which in the case of looking through the prism is not possible."[113]

The Field Goes Statistical: George Wilson's Researches on Colour Blindness

In the 1850s, Edinburgh natural philosopher George Wilson took several steps beyond Herschel's techniques by dramatically expanding his subject pool to 1,154 people and by turning attention to the *effects* of color blindness.[114] Abnormal color vision well suited Wilson's training as a physician, knowledge of color from his work in chemistry, and interest in technology while curator of the Industrial Museum of Scotland.[115] His studies included many previously reported cases, but also tests on new subjects from the local military, police, asylum, and his own classroom at the University of Edinburgh.[116] He hoped that his broad-ranging statistical analysis would organize the disparate experiences of color blindness. He sought a way around the impossibility of a "common language between those conscious of colour and those unconscious of it."[117] His work marked a watershed in color blindness studies since it decisively turned attention away from individual studies and toward a statistical understanding of the problem. His estimate of the frequency of color blindness among men—5.6 percent—was the first statistic ever given for the condition's prevalence. This figure startled and intrigued the natural philosophical and medical communities, who had previously seen cases only in isolation. Wilson's work made color blindness a national issue in a quantitative sense, just as Dalton's case had made color blindness a national issue in a symbolic sense.

Wilson's attention to statistics derived from a fresh concern that color blindness had its darker side: it could cause serious transportation accidents.[118] A letter John Tyndall had written to the *Athenaeum* planted the notion in Wilson's mind. Tyndall noted the fact that even railway engineers with normal vision might misread signals. Red signified that the train should stop, green to proceed slowly, and white to maintain full speed. Several laboratory and on-board experiments showed that the complementarily colored red and green lights seen together or in rapid succession often appeared white. Tyndall suggested that this problem might explain some railway accidents. Several months later, an intrigued Wilson wrote to the same journal to ask readers to send him details of their color vision problems, suspecting that color blindness among railway engineers might also be causing accidents. He and a colleague, mathematician Philip Kelland, had noticed among their students a ratio of one color-blind among thirty-seven or thirty-eight normal-sighted.[119] At his own expense, Wilson circulated this concern among the railway companies.[120] To his relief, he found that the Great Northern Railway Company already tested its employees for color blindness.[121] Brewster, convinced by Wilson's polemic, argued that color-blind persons should be barred from working in chemistry, pharmacy, food and beverage manufacturing, courts of justice, the army, or the navy.[122]

The threat of color blindness to national safety did not fully capture the British imagination until April 1876 when a spectacular railway accident occurred in Lagerlunda, Sweden.[123] Several well-publicized accidents in Britain, which seemed to have resulted from color blindness, further focused national attention on the problem.[124] Interest in color blindness increased as well during the last third of the nineteenth century, and the nature of the studies changed from individual cases (in many cases, self-studies) to statistical surveys. More than ever before, the safety of hundreds of people rested on the ability of one man to distinguish red from green. In 1890 the Royal Society convened a Committee on Colour Vision, led by Lord Rayleigh, to investigate the matter. After thirty meetings and tests on five hundred individuals, they concluded that the Board of Trade should institute mandatory tests for color blindness and color ignorance throughout the transportation industry. They recommended using a common set of wool swatches, so as to standardize the testing. Many branches of the military and transportation industries already did so.[125]

Through the increase in government vision testing and the Torres Strait anthropological study, late-nineteenth-century Britain inaugurated a new meaning for color blindness. Whereas earlier it had symbolized the triumph of discipline and judgment over physical imperfection in extraordinary men, at the end of the century it was the color blindness of the working classes and other races that garnered more attention. Now, correction of abnormal vision required quantitative knowledge rather than case studies, and standardized vision training rather than trust in individual judgment. It wasn't that the judgment of the top men of science no longer mattered at the end of the nineteenth century. Rather, their invention of standardized tools to manage subjectivity prevented them from needing to rely on the judgment of those "lower" classes and races who were presumed to lack it.

After Wilson, one of the most concerted efforts to deal with color blindness on a grand scale came from Liverpool physician Thomas Herbert Bickerton, whose evidence was used by the Royal Society's Committee on Colour Vision. For a full sixteen years (1885-1901), he conducted studies of the color-blind at the Liverpool Royal Infirmary. Skeins of wool in tow, Bickerton visited a vast number of the sort of institutions usually selected for Victorian statistical, reformist studies: schools, hospitals, asylums, and orphanages. He boasted that he could test sixty to eighty subjects an hour, and thanks to this efficiency and the promise of aiding the pursuit of safer public works, many of these institutions obliged the doctor.[126] In his tests of 3,087 sailors, Bickerton found that 105, or 3 percent, were color-blind. The seriousness of this statistic prompted him to scrupulously scan the *Times* for references to accidents caused by color-blind sailors, pilots, and railroad engineers. He found forty-two such references.

Not everyone agreed that color blindness should bar people from certain occupations. the color-blind engineer William Pole (about whom more later in the chapter), pointed out that "never, in a single instance, since railways have been in use, has an accident been traced to the mistaking of a red for a green night signal." Given the statistical prevalence of color blindness as reported by Wilson, he said, one would expect to be able to report several accidents due to red-green confusion, if the defect truly led inexorably to such accidents.[127] Perhaps railway engineers and sailors had better judgment than they had been given credit for. Despite the various interpretations of these statistics, the fact remained that statistics were replacing case histories because provincialisms of language and vision would longer do in a nation united by railways. Just after Wilson's book appeared to reify this trend, James Clerk Maxwell began publishing his own studies of color blindness, wherein he introduced another tool for standardization: quantitative distinctions of color.

Quantitative Precision Achieved: James and Katherine Clerk Maxwell

Maxwell, the man credited with firmly establishing Young's three-receptor theory of color vision, first entered the public discussion of this subject at George Wilson's behest.[128] In an 1855 letter to Wilson, published as an appendix to the *Researches,* Maxwell established himself as part of the new standardized approach to vision. Sensations, he argued, while fundamental to perception, were too simple to be analyzed: "If we attempt to discover them, we must do so by artificial means; and our reasonings on them must be guided by some theory." Understanding vision, whether normal or abnormal, required the same theory-guided, instrument-recorded, precision techniques that had come to define the rest of natural philosophy. Those were the conditions for achieving "a science of sensible colour independent of individual peculiarities."[129] Using this approach had made it clear to Maxwell that Young's theory had been both empirically sound and mathematically elegant. Namely, all color vision resulted from three different elements of unanalyzable sensation, which he called intensity, hue, and tint. Young, he argued, had identified correctly red, green, and blue as capable of producing all possible tints, but Maxwell stressed that any three spectral tints could be chosen, so long as they combined to produce white. The eye and mind actively combined these sensations into the variety of complex colors that we perceive.

Maxwell constructed a geometric figure to represent the fields of normal and color-blind vision (see Figure 3.3). He found that if he plotted on his triangle those colors that color-blind subjects confused, parallel lines could be drawn through each set of confused colors. In turn, he could then write simple equations to predict which further colors his color-blind subjects would report as equivalent. This, finally, presented Maxwell with a theoretical conclusion about color blindness:

> The mathematical expression of the difference between Colour-Blind and ordinary vision is, that colour to the former is a function of two independent variables, but an ordinary eye, of three; and that of the relation of the two kinds of vision is not arbitrary, but indicates the absence of a determinate sensation, depending perhaps upon some undiscovered structure or organic arrangement, which forms one-third of the apparatus by which we receive sensations of colour.[130]

Maxwell, in short, closed the chapter on color blindness as a problem of provincialism. He established differences in color vision as predictable, precise quantities. He thereby rendered obsolete the case history method, and organized subjective accounts under a set of methods, diagrams, and equations that his peers deemed universally applicable.

As with Dalton and Troughton before him, Maxwell's victory in physiological optics paralleled the broader trajectory of his career. Though he began his

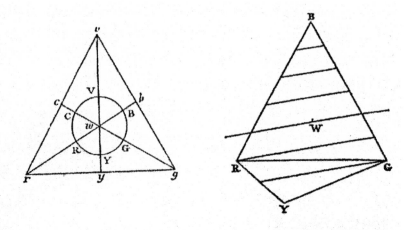

FIGURE 3.3
Maxwell's Color Triangles
The first triangle represents the field of normal vision, where v = violet, r = red, g = green,
w = white. By varying the amount of each of these colors on the top, he could produce sensations
all over the triangle. One could think of the three basic tints as variable masses, he said, and every
color as a center of gravity between different levels of mass. Maxwell could also predict the
response of the normal observer based on the numerical values of each color used on the top. The
second triangle represents the field of color-blind vision. Each parallel line runs through a set of
colors that the color-blind subject judges to be the same. Again, a key virtue of the diagram was its
predictive value. Source of illustration: Maxwell, "On the Theory of Colours."

life as the heir to a Scottish country laird, he ended it as a national hero.
Maxwell's first biographers, Lewis Campbell and William Garnett, described
their friend's father as having a "good deal of Presbyterian inertia in his compo-
sition," and as "one of a race in whom strong individuality had occasionally
verged on eccentricity." John Clerk Maxwell's "idiosyncracy, as has been said,
was well suited for a country life." Likewise, his son "might have been taken, by
a careless observer, for a country gentleman, or rather, to be more accurate, for a
north country laird. A keener eye would have seen, however, that the man must
be a student of some sort, and one of more than ordinary intelligence." His
friends admired how his "marvellous eye for harmony of colour" appeared in his
"plain and neat" dress.[131] Just like Dalton, Maxwell fulfilled the heroic Victorian
role of the provincial lad turned genius with universalizing talents.

Maxwell's interest in color had long predated his first paper on the subject. In fact, it dated back to his childhood fascination with his mother's and

aunt's knitting, which "early sent him inventing curious patterns and harmonising colours," and knitting garments himself.[132] Interestingly, where Dalton's encounter with his mother and knitted stockings led him to question her judgment, Maxwell's own experience only sparked his creativity. He began scientific work on color vision early in his career while serving as an assistant in James Forbes's laboratory at the University of Edinburgh. The subject had already received a good deal of attention from Edinburgh natural philosophers, including Forbes, Brewster, Wilson, physician William Swan, and David Ramsay Hay, decorative painter to the queen and author of several works on color classification and aesthetics. Hay believed that Britain needed to codify laws about the aesthetic contrast and harmony of colors in order to improve the country's refinement.[133] He therefore wholeheartedly supported Maxwell's work, and provided him and Forbes with tinted papers and tiles. Forbes and Maxwell developed a set of experiments that the latter would continue to pursue after he had left the University of Edinburgh: they spun different combinations of colors on a revolving disk to determine their combined effect on vision. Though the papers for this top were painted, the effect of the spinning top equaled what would have been achieved with spectral light.[134]

After his 1854 graduation from Cambridge, Maxwell returned to the study of color vision. In November of that year he tested ten observers with his color top, and published an extensive report on his findings, often considered the first modern study of color vision, for the methodological reasons cited above.[135] Maxwell's results gave credence to another aspect of Young's theory, namely that three systems of nerves conveyed color information to the brain. Maxwell explained that those nerves most sensitive to red sensation also conveyed small quantities of blue and green sensation. He illustrated this concept through photography. If one photographed a scene through a red lens, then a green, then a blue, and used a magic lantern to superimpose the resulting images on a screen, a full-color image of the original scene would appear. This indirect evidence for Young's hypothesis allowed Maxwell to explain color blindness without having the anatomical evidence. Using the same methods as before and four color-blind men (only two of whom produced reliable results), Maxwell tested and confirmed his hypothesis that red-green blindness stems from a lack of red-sensitivity, as Herschel had suggested. He found that *all* the colors visible to these subjects could be produced by combining only green, blue, white, and black.[136]

Through these experiments, Maxwell eventually arrived at a simple equation which described the missing sensation in a color-blind individual:

$$0.19G + 0.05B + 0.76Bk = 1.00R$$

where G = green, B = blue, Bk = black, and R = red. The equation was arithmetically simple, but methodologically important. Attaching a numerical value to sensations made it compatible with the psychophysical studies of "just noticeable differences" emerging from German laboratories. To Maxwell, this quantification at least reduced, even if it did not eliminate, the problem of communicating idiosyncratic visual experiences. The observer still had to communicate his color perception to the operator, though.[137] After testing his former student James Simpson, Maxwell felt the need to place quotes around Simpson's responses of "yellow" and "blue," in order to indicate their status as subjective accounts.[138]

From her 1858 marriage to James Clerk Maxwell, Katherine Mary Dewar contributed steadily to his research.[139] In Katherine, James found not only a worthy companion and assistant, but also another (mostly) normal eye against which abnormal color vision could be compared (see Figure 3.4). The couple faced a problem common to nineteenth-century experimentation: they had limited access to research subjects.[140] Acquiring subjects was not the only problem: the Maxwells also had to concern themselves with training them. Just as Wilhelm Wundt later trusted only highly trained observers to produce reliable psychophysical data, so too did the Maxwells instruct their subjects to judge color matches properly. They found that even "normal" observers gave significantly different responses to different colors. They assumed that the discrepancy resulted more from the observer's imprecision rather than any real physiological or psychological differences in perception.[141] They clearly wished to avoid the kind of language idiosyncrasies that had plagued earlier studies. To ensure the best results, they used their most trained subjects—themselves—as the "normal" observers.[142]

Another important experimental contribution that Maxwell made was to design a "color box" in 1858 to replace the top (see Figure 3.5). The new instrument allowed the Maxwells to avoid the discrepancies in dyes and papers that they encountered with the top. In early 1860, Edinburgh optician J. M. Bryson built a portable version, allowing the Maxwells to travel to subjects, and further collapse any problems of provincialism.[143] Using themselves as the "normal" subjects, the couple again showed four color-blind people a long series of color

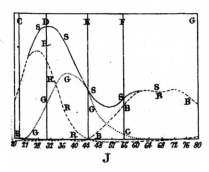
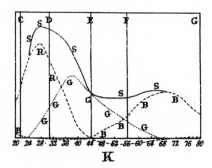

FIGURE 3.4
A Discrepancy between the "Normal" Observers
Depicting his own color vision in graph "J" and his wife Katherine's in graph "K," James Clerk
Maxwell showed how a slight anatomical difference affected their sensitivities to the various
parts of the spectrum. Maxwell suspected that his wife lacked the spot on the foramen centrale
that in most people absorbed much of the light betwen E and F, or the green portion of the
spectrum. Source of illustration: Maxwell, "On the Theory of Compound Colours."

combinations through the color box. They recruited the help of Simpson,
William Pole, Rev. W. C. Mathison, and a fourth subject whose identity is unknown. Each of these individuals was chosen not simply for his color blindness, but also for his otherwise superior observational skills.[144]

We can summarize Maxwell's major conclusions as follows: first, the normal human eye usually estimates the similarity of two colors with great accuracy. Second, the agreement of two colors indicated their common effect on the
normal eye, not the physical or "real" identity of those two colors. Finally, different observers' eyes vary in accuracy, but agree so much that one cannot doubt
that all normal eyes obey the same laws of color vision.[145] He also set a new
standard for precision and universalizable knowledge in color research. What
Dalton achieved biographically and symbolically, Maxwell achieved by revolu-
tionizing how the sciences managed subjective color vision. Without ample scientific training, subsequent students of color blindness sometimes found the
new methodological standards daunting to replicate. This indicates again how
hierarchical natural philosophy's system of management had become.

Struggling with the New Methodological Standards: William Pole

When civil engineer and organist William Pole decided to study his own color
blindness in the 1850s, the Maxwells provided procedural models, while Herschel

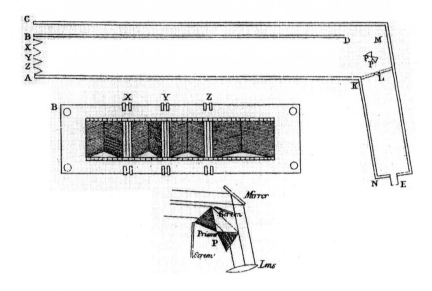

FIGURE 3.5
Maxwell's Original Color Box, 1858
The color box was about five feet long and seven inches across. Light entered the box at BC, reflected off mirror M, and passed through lens L. Light also entered through AB, to which were attached six brass sliders, as shown in the middle illustration. Moving the sliders enabled the experimenter to line up slits X, Y, and Z with any three portions of the pure spectrum. The instrument allowed precise readings of the slits' positions and widths. The three beams of colored light would be refracted through prisms P and P', and proceed through L. By placing her eye at E, the observer saw the resultant light of XYZ side-by-side with the light from BC, and could compare the two. The experimenter could adjust the white light from BC and the colored lights from XYZ to produce every visible color. Source of illustration: Maxwell, "On the Theory of Compound Colors."

helped him perform the initial experiments and commented on Pole's report to the Royal Society.[146] Pole's case was especially important, for his strictly blue and yellow sensations seemed to confirm the dichromic theory Herschel and Maxwell had outlined. But Herschel and others distrusted Pole's confidence that those afflicted with color blindness were always best suited to study it. Pole argued that the color-blind could not only report firsthand experience, but that they also had a richer understanding of the experiences of other color-blind observers, even those whose abnormality was slightly different. He did acquiesce, though, that only a minority of the already small pool of color-blind persons had the intellectual training necessary to give an accurate accounting of their visual experiences, "for it is important to consider that it is by no means an easy thing for a colour-blind person to explain his impressions to a normal-eyed one in language free from misapprehension." He observed that even those with normal vision frequently could not agree on what to call a given color.[147] Pole's case deserves attention because his awkward and patchy use of the new standardized methods seemed inadequate to his better-positioned contemporaries. Where his work would have been perfectly appropriate earlier in the century, it paled next to Maxwell's dazzlingly quantitative and totalizing conclusions.

Following the Maxwells' experimental design, Pole employed the services of the Edinburgh optician Bryson to construct a color top. He then found the closest match between colors produced by the top and those on Chevreul's color circle. Using these standardized tools, Pole described at length his experience of the spectrum, including the exact values of gray that to him appeared the same as various values of green. If put to the same test, he conjectured, the color-blind in general would reveal themselves to have an infinitely varied disorder. Pole began to confirm this when he found in studying several others that their experiences were qualitatively similar, but that the numerical color values were slightly different. Firmer proof of this conjecture, he said, would only come from further study.

The Royal Society expressed interest in the paper, but encouraged Pole to undergo more original research. Reviewer Charles Wheatstone admired Pole's clarity and claims to firsthand experience, but found his analysis unoriginal. Wheatstone argued that his case did not seem readily distinguishable from Troughton's or Dalton's, and therefore probably did not merit separate publication unless in a popular journal. He also found Pole lacking in the expert knowledge by then required to conduct a sophisticated investigation of color blindness: "The results would probably have been more satisfactory if the author instead of employing the printed tinted scales of Chevreul had used with Herschel and Seebeck the well defined and easily reproducible colours of polarized light." Neither the novelty of personal experience nor outmoded procedures could be tolerated any longer in nationally sanctioned studies of vision.[148]

Ultimately, however, the Royal Society heard Pole's paper at an 1859 meeting. Alerted to the possibility of training a good observer, several members began advising Pole on how to conduct further research and revise the paper for publication. G. G. Stokes, for instance, recommended that Pole include the name of his color top's manufacturer "because the value of your colour top exper[imen]ts depends in some measure upon their being reproducible." Later that year, Pole returned to the Royal Society so that Stokes could test his ideas on whether the color-blind experienced different hues of red simply as gradations of gray. In continuing his own investigations, Pole tested "3 or 4" other color-blind subjects. He found that color sensations varied not only with the observer, but also with the same observer at different times. He could not determine whether slight variations in lighting conditions or a *"really variable kind of impressibility of the eye"* caused these discrepancies, though he suspected the latter.[149]

Once again, Pole's paper went under consideration at the *Philosophical Transactions*. This time the lot fell to Herschel, whose review was mixed. He agreed that Pole's scientific attainments ultimately validated his testimony, but they did not quite qualify him to theorize on *causes* in this field of science. Herschel corrected Pole's error in positing red, yellow, and blue as primary colors, a mistake common to "artists, dyers," and others whose experience of optics was limited to pigments and amateur involvement in photography and astronomy. These comments further illustrate how rapidly vision studies were changing. A mere two decades before, a respected natural philosopher such as Herschel could and did still state that red, yellow, and blue were the primary colors of light, and Brewster always maintained this. Even Helmholtz did not consistently distinguish additive from subtractive color mixing.[150]

By the 1840s, however, Herschel argued that red, violet, and green could also be used as spectral primaries, and that receptors in the eye probably mirrored this tripartite division. If Pole was blind to red, he could not continue to see yellow and blue, since in the Young-Maxwell-Helmholtz theory yellow was a mix of spectral red and green, and could not be seen by someone blind to red. This discrepancy of views persisted despite the fact that Herschel and Stokes guided much of the research. While they were happy to let Pole report his own empirical findings as a sort of privilege, the natural philosophers central to the Royal Society did not view him as a legitimate engineer of optical theories. In fact, halfway through his review, Herschel began speaking of Pole as if he were a research subject rather than a theorist in his own right.[151] Pole encouraged this relegation to inferior status in his own writing by making it clear that he conceived of his work as a mere explication of abnormality rather than a grander attempt to understand normal vision. "All I wish to do is to give a statement, as explicit and accurate as I can," he wrote, "of the facts connected with the sensations of myself and others similarly situated, leaving these facts [of normal vision] to be discussed by others more competent to deal with them." Pole never lost this sense that he was a fact-collector who had no real business among the philosophers.[152]

Still, he persisted. Over the next few decades, the engineer's continued work on color blindness displayed a growing sensitivity to the importance of standardization.[153] With instructions from Maxwell, he ordered a new color top from Elliott's, an eminent mathematical instrument shop on the Strand in London. He commissioned the colored papers for the top from Messrs. de La Rue, the first shop to produce them commercially. By then, Pole knew the importance of emphasizing his abnormality as a quantitative key to normal color vision, rather than presenting his personal observations simply as a strange case history. For instance, in 1860, he sent colored cards to Stokes and asked that he select the "most definite & perfect colours." Pole certainly could not select them himself, and since he wished to "establish the instrument permanently on a better footing I should wish them selected by some eminent authority."[154]

By 1864, Herschel was sufficiently impressed with Pole's rigor to hand over the silks and data from his test of Dalton's color vision. The aging Herschel encouraged Pole to record his own responses to the silks. But this time, Pole added an element of standardization unavailable during Dalton's original test: he found a match for each silk among Chevreul's color scales. By finding the closest color match among these standard set of hues, Pole came much closer to an objective indicator of the difference between normal and color-blind vision (see Table 3.1).[155]

Contained within this table are a number of by-then conventional strategies for generalizing the provincial knowledge of the color-blind. The first section of "single samples" utilized the original method of having the subject attach a free-form, verbal description to a colored object. The responses become more generalizable in the second section, "samples matched together," which appears to be modeled on the second round of tests that Herschel performed with Troughton. As Pole put it, "[M]atching shows the nature of the vision independently of nomenclature, which is not always to be depended on." Finally, Pole's column achieves a further level of standardization in two ways. Because of the references to Chevreul's circle, other observers could repeat the experiment without fear that their individual sense of "light blue" would obscure the real differences or similarities between their vision and that of other observers. Furthermore, by incorporating Herschel's theory of dichromism into the experiment itself, Pole thought he gave a truer account of how the color-blind perceive colors. Dalton and many other color-blind persons had claimed the ability to see red, but Pole argued that he had simply learned to label certain yellowish hues as red because others did so. It was only after his long investigation of the ramifications of Herschel's theory that Pole could conclude "that

"Normal" description of the colours	Dalton's description	Description by "another colour-blind person" [Pole]
I. Single samples		
Crimson	Reddish brown	Yellow, 18
Brown	Yellowish brown	Yellow, 18
Red violet	Red and blue; latter prevails	Gray, 16
Gray	Slate blue, brighter than the foregoing sample, but nearly allied to it	Gray, 13
Blue black	Blue black	Blue, 20
Black, pure	Brown black	Black, no colour . . .
II. Samples matched together		
A. Orange	Orange yellow; alike, and	Yellow, 14
Yellow green	nearly the colour of gold	
B. Yellowish pink (salmon)	Alike on first glance, but there	Yellow, 13, with less colour
Yellow green	is a shade of difference	Yellow, 16
C. Green	Brown and nearly alike, first	Yellow, 19 ½
Red	rather brighter than 2nd & 3rd	Yellow, 17
Brown		Yellow, 19
D. Blue, pure	Dark blue, nearly alike, first	Blue, 13, or 14, but colour
Violet	rather more vivid	more intense

TABLE 3.1
A Sample of Dalton's Qualitative versus Pole's Quantitative Descriptions of Color
Based on a table that appeared in Pole, "Daltonism," 825–26.

what I took to be red was merely a modification of one of my other sensations; and if, before I found this out, I had been interrogated on the subject, I should have declared I saw red as a separate colour, just as Dalton did."[156]

CONCLUSION

Not until the next century would Pole's views be vindicated. In her landmark studies of color vision, Christine Ladd-Franklin attacked the many supporters of the Young-Maxwell-Helmholtz theory who had neglected the arbitrariness of red, green, and blue as primaries. Both Herschel and Helmholtz had cautioned that those colors had been chosen purely for demonstration purposes—

in theory at least, other combinations could produce white and the full range of colors. Regardless, most vision scholars in practice reified red, green, and blue as *factual* primaries by asserting that visual receptors followed this tripartite division.[157] Ladd-Franklin resurrected Pole as someone who had not allowed theory to overwhelm the facts of the matter. Pole "convinced himself that his own sensations were blue and yellow," she wrote, "and he should have convinced all the world as well if the world had been open to reason — if it had not been preoccupied with a theory."[158]

If Ladd-Franklin was right, and Pole was a kind of provincial martyr to the snobbery of late-nineteenth-century physics, his contemporaries at least hoped to have minimized such prejudice. A strong reformist movement within natural philosophy had sought to ensure that any reasonably educated, intelligent individual could have access to the kind of standard materials and techniques that would make their local knowledge generalizable, and therefore valuable. However, the problem was that in reality few could access the ever-increasing training that good observations required. Subsequently, the distance between observer and subject grew, thus buying the authority of objectivity at the expense of the authority of personal experience. The scientist's personal experience was no longer necessary if standardized protocols and training procedures for observers made others' experiences transparent.[159]

Earlier in the century, reformers took a first step in this direction by zeroing in on the example of John Dalton, who gave them a heroic icon for the very epistemological and institutional network that they sought to establish. His transcendence of his various limitations allowed him to conceive one of the most important scientific theories of his day, the atomic theory, and indicated that the reform project could succeed on a grander scale. Dalton became a respected member of the natural philosophical community despite the provincial tendencies of his Quakerism, rural upbringing, and color blindness. He had done so by taking advantage of experimental and theoretical protocols. Figures such as Dalton — or Prince Albert, who overcame aristocratic "color blindness" against the technical arts, or Edward Troughton, who constructed beautifully precise instruments despite his flawed vision — provided exemplars of extraordinary ability. For the humbler man, the standardization of observational techniques in the next several decades theoretically meant that one did not need such incredible moral character to achieve transcendence over the local.

This was precisely what scientific and political reformers in the early nine-teenth century had hoped to achieve. Reform efforts in Britain aimed to use metropolitan resources to expand the experience of those in the provinces, thereby creating a truly national, and rational, government. This project mat-tered for a number of reasons. In the case of color blindness, overcoming this myopia would help democratize the cultivation of taste, and would lead to greater safety in transportation. Mechanics' Institutes and inexpensive scien-tific literature were meant to help achieve this by nationalizing natural knowl-edge. The proliferation of universities ensured that education became more ac-cessible to those who were not Anglican, wealthy, or male. Nationwide studies of health and labor conditions would enable the development of national pol-icy to improve life, especially in the provincial towns. Ironically and impor-tantly, this triumph worked to Pole's disfavor. Herschel's, Wilson's, Maxwell's and others' efforts to standardize the study of color vision had made Pole's idio-syncratic case history obsolete. The very tools that were supposed to democra-tize the sciences frequently reinforced the inductive hierarchy described in the previous chapter.

Provincial curiosities may still have enjoyed some amusement value in the 1830s, but they were signs of a time that reformers sought to leave behind. Ab-normality celebrated for its own sake had no real place in a nation or a science that sought to redefine abnormality as correctable variations from a mean. By midcentury natural philosophers had determined that certain differences in physiology, religion, aesthetic taste, morals, and color vision were simply irre-ducible to one common human nature. But by the same token, the very nation-alization through statistics and quantification that allowed this conclusion, also enabled the *control* of these variations. Abnormal vision was just another con-trollable variable, no longer as threatening to the epistemological certainty of the natural philosophers who had it.

FOUR

Mental Governance and Hemiopsy

The study of abstract and metaphysical theories, about which an ingenious person may reason plausibly, without ever arriving at a determinate conclusion, will often excite the organs of the reflecting faculties into sleepless and uncontrollable activity and disease; and, in fact, it often did so in former times, when men devoted themselves more to abstract pursuits, and mingled less with the world.

— Andrew Combe, *Observations on Mental Derangement*

ANYONE WHO HAS SUFFERED A MIGRAINE has probably watched spots, lines, or other geometric shapes radiate through their field of vision. Today, physicians call these visual effects scotomas. In Britain's early industrial period a number of prominent natural philosophers experienced the same phenomena, which they deemed the result of a nervous disorder called hemiopsy. Typically, a hemionope (someone with hemiopsy) would experience a sudden blind spot in one or both eyes; this spot would expand and become surrounded by jagged, colored lines; a headache often followed.[1]

Like the other nervous disorders considered in this book, hemiopsy led natural philosophers to discuss explicitly their views about proper management of the body and the nation. More than other nervous problems, hemiopsy had a particular association with overwork. Natural philosophers sought prevention and cure for hemiopsy in the new theories of work they were developing to improve mechanical efficiency. Hemiopsy especially encouraged natural philosophers to connect their own bodies and work with industrialization and the divisions of labor discussed in chapter 2.

Natural philosophers' discussions of hemiopsy focused on two specific features of the body/scientific practice relationship: first, the efficient management of the natural philosopher's body, and second, the relationship between mind and body. Clarifying these issues served a common goal, namely, determining how one could obtain reliable information about the world when one had an abnormal body. How could such a body be suffered to degrade the quality of the precise and accurate instruments and methods on which the era prided itself?

The subjects of this chapter gave the following answer: the natural philosopher must manage an abnormal body just as he would any other technological device in his service. The body, like the steam engine, attracted the attention of natural philosophers who sought dynamic efficiency. The mind, ambiguously positioned alternately inside and outside the machine, acted as a regulatory governor. If, as Norton Wise and Crosbie Smith argue, "political economy helped to solve problems in natural philosophy," this chapter argues that it also helped solve problems in mental-moral philosophy.[2] The analogies repeatedly drawn between engines and bodies highlighted natural philosophers' confidence that the rational mind could manage virtually any mechanical inefficiency in the body. Natural philosophers placed so much confidence in the rational mind, in fact, that it virtually ensured the perpetual motion of the body. While many natural philosophers saw irrevocable directionality in engines, bodies, economies, and other thermodynamic systems, they saved a place outside these mechanical systems for the managers: God, the mind, and the factory supervisor.

In the Enlightenment, hemiopsy resulted when one's body fell out of balance. From the early nineteenth century—even before William Thomson fully elaborated the concept of work in the mid-1840s—natural philosophers looked for explanations of hemiopsy in the level and direction of work (regulating power), rather than in the balance of mental and physical faculties. By the mid-nineteenth century, misgoverning the body-engine caused hemiopsy. In other words, inefficiency replaced imbalance as the ultimate pathology. Combatting inefficiency meant not only managing the body with the work-maximizing equations and techniques used for engines; it also meant trusting the extra-mechanical power of the mind to overcome physical problems. Bodily and social systems could not always right themselves. Early industrial natural philosophers argued that they needed actively to manage these systems in order to optimize their efficiency. Thus, once again, natural philosophers' nervous disorders ironically enabled them to demonstrate the broad power of their methods.

I will first discuss the turn-of-the-nineteenth-century shift from thinking about the body as a static object to be kept in balance, to thinking of the body as a dynamic engine whose efficiency the mind could ensure. I use some of the earliest writings on hemiopsy to illustrate these points. Then, I turn to four prominent hemionopes: William Hyde Wollaston, David Brewster, George Airy, and John Herschel. Chronologically ordered, these cases reveal in striking detail how natural philosophers began to connect hemiopsy with industrial management. By 1870, when Airy's son Hubert synthesized the extant scientific work on hemiopsy, the need to mentally manage the unruly body had become an obvious necessity in a world where machines were invaluable but unpredictable.

STATICS: THE TIMELESS STABILITY OF THE BODY

At the center of Enlightenment thought lay the belief in an orderly universe, a "balance of nature." This model took a wide variety of forms, but had three common features: "(1) an opposition of two forces, or two 'constant causes,' producing regular variations or oscillations about a mean, or equilibrium state; (2) irregular variations or fluctuations, also averaging about the equilibrium state, but produced by 'accidental' or 'disturbing' causes; and (3) optimization of the equilibrium as a result of variation."[3] Given that Enlightenment thinkers fully expected the human body and mind to adhere to natural laws, it should come as no surprise that we can discern these same features in mental-moral philosophy. For Enlightenment philosophers and physicians, the human mind had a normal state but could be disturbed out of its natural equilibrium by a variety of excesses. Both humoral and vital-force conceptions of the body equated disease with imbalance. Accordingly, the French physiologist François-Joseph-Victor Broussais identified pathology with physical excess or deficiency. Pierre-Jean-Georges Cabanis exhorted physicians to teach their patients mental habits that would restore physical and moral equilibrium to the human frame. Locke conceived of pleasure and pain (the roots of all action) as "different Constitutions of the Mind, sometimes occasioned by disorder in the Body, sometimes by Thoughts of the Mind." Hartley rested assured that this vacillation of passions, the tug-of-war between pleasure and pain, would ultimately result in humanity's virtue and happiness—and God ultimately guaranteed that balance. Hume also considered the passions to be subject to "a certain regular mechanism, which is susceptible of as accurate a disquisition, as the laws of motions, optics, hydrostatics, or any part of natural philosophy." Benjamin Franklin warned his fellow Americans to "avoid extremes. Forbear resenting injuries so much as you think they deserve."[4]

This understanding of human bodily and mental life as vacillating around a stable equilibrium merged well with the early principles of probability. Laplace tolerated probability as a necessary evil on the sure path to deterministic knowledge. By the early nineteenth century, however, this attitude had changed dramatically. Probabilism became recognized as a constitutive feature of many natural and social phenomena. In other words, the sciences needed probabilities not because their knowledge was imperfect, but because many phenomena did not follow deterministic laws. By extension, abnormal phenomena became regular features that occurred in semi-predictable fashion.[5] This altered view contributed to thinking of bodily and mental abnormality and the inefficiency they caused as unavoidable realities to be managed rather than eliminated or denied.[6]

But up to and during the Enlightenment, the varying readings of different bodies and instruments were interpreted as errors — errors of judgment (in reading the instrument or interpreting the sensation), of calibration, or of manufacture. Even in experimental settings, academic mechanical philosophers sought to eliminate fluctuations in the hopes of demonstrating abstracted Newtonian laws. Behind every accurate outcome lay an adumbrated but necessary human judgment. In Enlightenment thought, the will (divinely given or not) helped steer natural philosophers to the accurate reading, and militated against the fluctuating errors that resulted from human imperfection. In contrast, in the early nineteenth century, the management of bodily and instrumental error became the rallying cry of the proponents of analysis, those who sought to go about the "real" business of studying power and the most efficient ways to use it. The old "Cambridge reality" which was "defined in Newtonian terms as void and resistanceless" was gradually replaced with the engineering approach to mechanics, "modeled through friction and mechanical effect."[7] In a system where friction and other imperfections could not disappear, the will to work through them became even more crucial. We can see this shift over the history of studies of hemiopsy: accounts of the disorder as an imbalance were replaced during the nineteenth century by discussions of hemiopsy as a problem of efficiency, of work and waste.

Accidental Colors, Ocular Spectra, and Hemiopsy: Early Accounts

By the late nineteenth century, one strange visual disorder had attacked so many prominent natural philosophers, that a young Cambridge student named Hubert Airy decided to write his thesis about it. He began with his own experiences and those of his father, the Astronomer Royal George Airy. Through his father's exceptional connections, he also collected information from other prominent men of science: William Hyde Wollaston, François Arago, David Brewster, John Herschel, and Charles Wheatstone, among others. Hubert coined the term *transient hemiopsia* (literally, temporary half vision) to denote a related set of afflictions in these men.[9]

Consistent interest in conditions like hemiopsy had spanned more than a century between Buffon's initial notice of it in 1743 to Hubert Airy's 1870 presentation to the Royal Society. Enlightenment investigators lumped together hemionopic effects with all manner of subjective visions as "afterimages" or "accidental colors," the colored patterns one saw after staring at the sun, prodding the eye, or otherwise inducing visual distress. Newton's experiences with afterimages provided the entrée for subsequent philosophers. Newton's "phantasm" resulted from overzealous experimenting with sun-induced afterimages. It took three days in a dark room to exorcise the visions, and years later they still haunted him. They often deprived him of sleep, and may even "have driven him to the verge of distraction."[9] Describing these events to John Locke, Newton threw up his hands as to the "cause of this phantasm," and confessed that understanding the complex interaction of perception and imagination in this case was "too hard a knot for me to untie."[10]

More persistent early investigators considered the phenomena "accidental," a temporary fluctuation from an otherwise normal state of vision.[11] A typical eighteenth-century explanation of these images suggested that the fluctuating contrast colors corresponded to a temporary imbalance in the nervous mechanism, which gradually righted itself. For his medical thesis at the University of Leyden, Robert Waring Darwin confirmed this. Using the research on accidental colors already performed by his father, Erasmus, the young physician argued that the retina was in "an active not in a passive state during the existence of these ocular spectra." Following the original excitation of the retina came a series of progressively diminishing retinal spasms, until finally the afterimages ceased, much like a pendulum gradually reaching its center.[12]

The term Darwin used to describe these visual phenomena, *ocular spectra*, remained popular throughout the nineteenth century. By adopting this term to indicate this range of visual effects, investigators connoted the uncanny. "Spectrum" and "spectre" are etymologically related terms, and suggest apparitions. Seers of apparitions traditionally had been viewed as either divinely inspired or mad. By the late eighteenth century, however, British natural philosophers tried to claim some apparitions for the world of secular sanity. While the medical profession generally continued to use hallucinatory episodes as the primary indicator of insanity, a growing number of natural philosophers—and later in the century, psychologists—would argue that many visions were simply the product of a mistake in judgment and thus quite ordinary (see the next chapter).[13] This signaled a subtle but important change from earlier theories of "accidental colors," which existed only as unusual disturbances. By contrast, in their new incarnation as "ocular spectra" and in the particular form of "hemiopsy," these visions were understood as a regular feature of the energetic life, and particularly the life of the natural philosopher. It was in this context that hemiopsy became an object of study. Hemiopsy was a particularly extreme version of ocular spectra, supposed to result from the extraordinary stress *regularly* endured by active men, and managed by the superiority of that man's intellect.

The reconception of ocular spectra (and especially hemiopsy) as regular rather than irregular features of the active body was made possible by natural philosophy's shift in models from the balance to the engine. As with the specific case of hemiopsy, so with the broader conception of human capacity: the Enlightenment understanding of abnormality as mere fluctuation from the stable norm gave way to an acceptance and management of inefficiency as a regular feature of human life.

On Governors

The "dramatic transformation . . . from balancing models of natural systems to engine models" that occurred in political economy, astronomy, geology, natural philosophy, mathematics, and engineering, also took place in mental-moral philosophy.[14] Natural philosophers who were otherwise active in optics, astronomy, electricity, and so on, found the steam engine compelling for understanding their own bodies and minds. We can take investigations of hemiopsy in the nineteenth century as an indication of how natural philosophers recast the body (including the brain) as a steam engine and the mind as the machine-tender or governor which regulated its output. The steam engine as an explicit model for other natural processes did not achieve its full popularity until after midcentury.[15] Even before the appearance of thermodynamics, however, we can find frequent comparisons of the body to an engine, and more generally to a mechanical process that, like the factory, required careful management to achieve the optimum efficiency of input and output. And while efficiency was just as crucial in nonindustrialized portions of the economy, the science and technology of steam production especially galvanized the imagination of natural philosophers around the issue of efficiency.[16]

Steam Engine Physiology

To begin uncovering this fascination with the body as an engine or factory we need look no farther than the best-known contemporary paean to industrial production, Andrew Ure's 1835 *Philosophy of Manufactures*. Ure, a University of Glasgow professor and fellow of the Royal Society, concluded after numerous factory tours that industrial manufacturing satisfied all reasonable scientific, moral, and commercial expectations. He deemed ill-founded the complaints about the well-being of factory workers (or, as he called them, "inmates"), and urged workers to conform to their masters' designs. Manufactures, he wrote, have

> three organic systems: the mechanical, the moral, and the commercial, which may not unaptly be compared to the muscular, the nervous, and the sanguiferous systems of an animal. They have also three interests to subserve, that of the operative, the master, and the state, and must seek their perfection in the due development and administration of each. The mechanical being should always be subordinated to the moral constitution, and both should co-operate to the commercial efficiency. Three distinct powers concur to their vitality, — labour, science, capital; the first destined to move, the second to direct, and the third to sustain. When the whole are in harmony, they form a body qualified to discharge its manifold functions by an intrinsic self-governing agency, like those of organic life.[17]

Ure wanted this complex analogy between factory and body, schematized below to serve several ends (see Figure 4.1). First, he wished to naturalize the industrial economy, thus making moral arguments against it seem futile. Second, and especially important for my argument, Ure sought to establish science as the nerve center of the industrial nation's body. He appears to mean "science" in its broadest sense, namely, the natural laws of economics, morals, and nature. If workers/muscles labored, and the state/blood vessels disseminated the products of that labor, then the scientific master/nerves ensured the rationality of this process.

Scientific managers organized and directed labor so that as much as possible became available for the state's commercial enterprise. Without the efficient cooperation of each element, this manufacturing "organism" would die.[18]

Ure's optimism about the morality of manufactures competed with a conservative apprehension about the corrosive effects of industry. Poet laureate Robert Southey, for example, reviled mechanical industry as "a fungous excrescence from the body politic." This tumorous growth had grown so large and had infiltrated so deeply that its host could neither survive its presence nor its excision. In Southey's version of the body-factory analogy, the nerves and vessels did not direct and disseminate healthy functioning. They spread contagion.[19] Ure's and Southey's voices joined thousands of others who debated the "machinery question" in early-nineteenth-century Britain. Not only in the pages of books and journals, but also through social movements and machine breaking, debate raged on the morality and utility of mechanizing the British economy. We must therefore understand any comparison of bodies to machines by British natural philosophers in light of these debates. The very people who cast the body as a machine were the same reformist members of the "scientific movement" who supported strong ties between political economy, the state, industrialization, and the sciences. Even those such as Carlyle who famously deplored the ubiquitous application of mechanical metaphors, usually rejected them only so far as they reduced humans to deterministic behavior. He found many sympathetic ears, since determinism had little cachet even among supporters of the inductivist hierarchy described in chapter 2.[20]

British natural philosophy in the early nineteenth century faced the profound challenge of answering honestly Romantic concerns about self-determination while retaining some of the Enlightenment's optimism about a

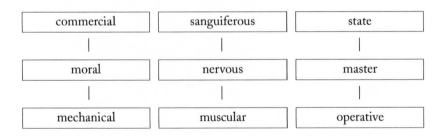

FIGURE 4.1
Andrew Ure's Conception of the Industrial Body Politic

lawful universe. Looking to the French school, many British physicians and physiologists strove to make the living, changing organism the preferred object of study while still focusing on measurable characteristics. In their drive to understand living function, British natural philosophers and physicians continued to use mechanistic tools to comprehend the body. To accommodate the prevalent moderate materialism, the body and the steam engine it resembled were always more than the sums of their parts. The mind as a governing entity continued to appear in nineteenth-century mental-moral philosophies, and this position was made compatible with—though not reducible to—a mechanical view of the body. Sara Coleridge, for example, viewed herself as no radical mechanist when she asserted that "life is the steam of the corporeal engine; the soul is the engineer who makes use of the steam-quickened engine."[21]

That engineer not only made use of the body-engine, but also kept it in working order. In most cases the body's fine design made the engineer's interference unnecessary. For example, when Brewster explained the perfectly normal phenomenon of accidental colors—such as the green spot one sees when one closes one's eyes after staring at a red spot—he argued simply that the eye was "strongly excited by the continued action of the red rays" in simple, mechanical fashion.[22] The mind's faculty of judgment was utterly unnecessary in such a simple process, just as a manager did not need to oversee continually the simpler aspects of a machine's operation. If one wanted to explain and understand this process, however, the reliable judgment of the manager had to be called into play, and the natural philosopher's physical life was never, therefore, *just* mechanical. In his *Economy of Machinery and Manufactures,* Babbage had emphasized "uniformity and steadiness" as crucial to keeping a complex machine running smoothly.[23] Similarly, he wrote after his own run for a North London Parliament seat that the House of Lords "is to the political what the fly-wheel [another power regulator] is to the mechanical engine. It ought to represent the average but not the extreme opinions of the people."[24] Like the governor on a steam engine Babbage admired so much, the philosopher's and the parliamentarian's attention served the dual function of deeply understanding function and managing abnormality.

The point of standardizing procedures in nineteenth-century natural philosophy, again, was not to replace entirely personal judgment, but to discipline it.[25] This discipline was crucial, because bodies and engines seemed equally vulnerable to breakdown. Much of the work that led to thermodynamics, for example, was directed at avoiding abrupt changes in velocity in machines.[26] Several sensational steam boiler explosions in the first few decades of the nineteenth century dramatically highlighted the importance of vigilance. Even when the safety valve was left open, so much steam could form rapidly around the boiler's red-hot parts, that the boiler could not contain the pressure. After each explosion, the public often complained that boilers had been pushed beyond their capacities and were monitored by untrained lackeys. No wonder, then, that the English celebrated their countryman Richard Trevithick for developing a functioning, high-pressure steam engine, something James Watt had considered too dangerous to construct.

One solution to the problem of engine overload was to have a mechanical device such as the steam-engine governor regulate power. Governors kept a machine running at a uniform rate, thereby simultaneously maximizing efficiency and safety. The governor consisted of two heavy balls attached to a spindle, whose rate of rotation depended on that of the flywheel. The varied divergence and convergence of the two balls affected a set of levers that controlled the quantity of steam admitted to the cylinder. If the flywheel slowed, for instance, the balls moved together, thus opening the throttle valve and admitting more steam.[27] An admiring Babbage explained that "wherever the increased speed of the engine would lead to injurious or dangerous consequences, this is applied; and it is equally the regulator of the water-wheel which drive a spinning-jenny, or of the wind-mills which drain our fens."[28] An 1860s advertisement promised "no more explosions," thanks to a governor that "seems endowed with intelligence."[29]

Another regulatory device, the cataract, regulated a steam engine's strokes through a valve "adjustable at the will of the engine man." Whereas a cataract in the eye obstructed the passage of light and therefore vision, one in an engine could either obstruct or free its motion. In the previous century, Soho manufactory founder Matthew Boulton had invented the mercurial siphon gauge, a meter that displayed the level of vacuum in the condenser. Without this "outward and visible sign," he said, "it is impossible to judge of ye inward & spiritual grace." Boulton particularly enjoyed monitoring the grace of his machines by watching the mercury bob up and down in the gauge.[30] Supervising the health of a machine was a necessary, even an enjoyable, aspect of industrial management.

These concerns about regulating industrial power migrated into studies of the human body, the analogy becoming especially popular after midcentury. Lyon Playfair argued that the body was a machine fueled by food, "in the manner of any steam-engine," but it required the heart to act as the "superintending engineer who watches the working of the animal machine that physiologists call Life." The human body had the added benefit of the mind as a superior governor.[31] Though they agreed that the laws of energy conservation governed the body, both William Carpenter and Peter Guthrie Tait equally disdained materialists such as Huxley and Clifford who studied the body as though there were no mind, and spiritualists who committed the opposite error.[32] George Henry Lewes noted that while machines and organisms shared superficial similarities, an overzealous love for analysis could obscure some important differences. Unlike the machine, the organism consisted of "interdependent and primarily related parts" whose "integrity depends on the continued destruction and renovation of their substance." In other words, "waste is a condition of their vitality."[33]

But apart from the necessary waste from the digestive system and cell reproduction, there was still the problem of what we might call humans' external work and waste. Midcentury mental-moral philosophy abounded with a concern for maximizing human, and especially scientific, work. To achieve extended "intellectual operations" with minimum fatigue, Carpenter recommended a mix of methodical training for the mind, along with maintaining bodily health.[34] Helmholtz suggested that investigators stop worrying about the imperfection of their perceptual instruments and concentrate more on how they *used* those instruments. He thereby helped to turn the focus from the visual apparatus narrowly conceived to broader nervous functioning, which he likened to the telegraph systems then being installed throughout Britain (see chapter 6). He found another suitable model for human physiology in the steam engine. Helmholtz modeled his graphic methods for depicting muscle contractions on James Watt's indicator diagrams for measuring the work performed in a steam engine's cylinder.[35] Likewise, with his concept of correlation of forces, William Robert Grove sought to identify "scientific with social and economic progress" and to link "this progress to the idea of the machine as an instrument of social discipline and control." In his scheme, the experimental philosopher became like the factory manager.[36]

Not even the hyper-materialist Thomas Henry Huxley could quite resist the powerful image of the body-engine guided by a regulatory intelligence. Defining the liberally educated man for a popular audience, he said such a man would have been

> so trained in youth that his body is the ready servant of his will, and does with ease and pleasure all the work that, as a mechanism, it is capable of; whose intellect is a clear, cold, logic engine with all its parts of equal strength, and in smooth working order; ready, like a steam engine, to be turned to any kind of work, and

spin the gossamers as well as forge the anchors of the mind; whose mind is stored with a knowledge of the great and fundamental truths of Nature and the laws of her operations.[37]

For Huxley, such discipline turned men of science into models of liberal education and therefore national leaders. As we will see, this valorization of the well-managed natural philosopher began earlier in the nineteenth century, when men of science developed routines and practices to keep their body-engines in order.

Exercise versus Melancholy

The natural philosopher, whose mind was stored full with "the fundamental truths of Nature and the laws of her operations," had to take particular care not to run his own engine to the point of exhaustion. As noted in the first chapter, scholars had long been thought prone to melancholy. Since they did not engage in the physical exertion that preserved the working classes from such nervous conditions, intellectuals and other privileged classes sought another outlet. Exercise — not to produce goods or services, but for the pure sake of exertion — became more and more popular during the nineteenth century as a means for lifting melancholic spirits.[38] For example, in his lectures at the Central London Ophthalmic Hospital, Alfred Smee told his fellow surgeons to instruct their scholarly patients to vary their activity for the sake of their health. Short-sightedness almost inevitably resulted from uninterrupted poring over books, and should be tempered with time in the open air. "The study of morbid vision," he said, "demonstrates the necessity of dividing our attention between the books of man and works of nature."[39]

When we examine the biographies of natural philosophers, we can believe that a doctor such as Smee never lacked for intellectual patients. For instance, the overwhelming spectacle and organization involved in the early meetings of the British Association for the Advancement of Science taxed a few participants into a stupor: assistant secretary John Phillips had a nervous breakdown after the 1837 meeting in Liverpool. He found solace and renewed health in boating and climbing. And, as noted in chapter 1, the Bristol meeting seems to have been the proximate cause for William Henry's suicide.[40] George Wilson met an untimely death after excessive exertion in the Highlands, exacerbated by a grueling lecturing and curatorial schedule at the University of Edinburgh. "His weak body," J. H. Balfour lamented, "seemed often to be sinking into the dust, while his noble spirit ignored its fetters, and seemed to rise above the feebleness of the flesh."[41] In the early 1840s, Michael Faraday also buckled under a rigorous workload. His symptoms are not entirely clear, but he suffered enough to abandon scientific work for two years. He had been maintaining a full schedule of public lecturing and private experiments, and some work with volatilized metals early in 1839 left him with severe eye inflammation. He told John Herschel that under those conditions he was "obliged to use these organs very cautiously & for but short periods." He found relief from the "mental strain" at the theater, on the beach in Brighton, and in Switzerland. Reflecting on Faraday's life several decades later, John Tyndall admired the great chemist's ability to regulate his energies: "This, at one period or another of their lives, seems to be the fate of most great investigators. They do not know the limits of their constitutional strength until they have transgressed them. It is, perhaps, right that they should transgress them, in order to ascertain where they lie. Faraday, however, though he went far towards it, did not push his transgression beyond his power of restitution."[42]

In 1851, John Couch Adams worried about his mentor, Cambridge Observatory director James Challis, who had retreated to the country while "getting over the effects of overwork."[43] Astronomy popularizer Agnes Mary Clerke also pushed herself to the verge of exhaustion. She complained that the pool of astronomical facts grew larger each year, "and the strain of keeping them under mental command becomes heavier." The work she had written just prior to making this remark, *Problems in Astrophysics,* had so exhausted her that she could only labor at it for half-hours at a time.[44] After the death of his wife, Charles Babbage received a consoling note from *Encyclopaedia Metropolitana* editor Edward Smedley, who was rapidly going deaf. Smedley "dread[ed] . . . that power of compression which the mind sometimes so fearfully exerts. . . . Time — and Time only, is our certain Physician under mental suffering."[45] In 1861, William Thomson bemoaned his fractured thigh, which prevented him from attending the first meeting of the British Association's Committee on Electrical Standards, whose actions proved so important to the future of the telegraphy industry. Though the fall did not injure what he called his "vis viva," the "constant uniformity of my position prevents me from getting any refreshment and makes anything like head work to be avoided."[46]

In fact, by the end of the century, Francis Galton deemed energy the "most important quality to favour" in eugenic schemes, and found scientists extraordinarily endowed with it. His own breakdown at Cambridge had made this conclusion quite personal. Galton interpreted his frightening experience through the by-then familiar model of body-as-engine / mind-as-governor:

> I suffered from intermittent pulse and a variety of brain symptoms of an alarming kind. A mill seemed to be working inside my head; I could not banish obsessing ideas; at times I could hardly read a book, and found it painful even to look at a printed page. . . . I had been much too zealous, had worked too irregularly and in too many directions, and had done myself serious harm. It was as though I had tried to make a steam-engine perform more work than it was constructed for, by tampering with its safety valve and thereby straining its mechanism. Happily, the human body may sometimes repair itself, which the steam-engine cannot.[47]

Galton's active mind gave him the capacity to manage his body by occasionally releasing the pressure. In fact, he became so convinced of his mind's directive power, that at one point he nearly asphyxiated himself. In an experimental attempt to "subjugate the body by the spirit, and . . . determined that my will should replace automatism," Galton made his breathing dependent on his will, and had some difficulty rendering it automatic again.

For most nineteenth-century natural philosophers, though, the crucial point was that only the superior intellect could guide the hardworking body to maximum efficiency. This context is essential for making sense of hemiopsy and its implications for scientific work, for the natural philosopher concerned himself with the optimization of not only heat economy and thermal efficiency in the steam engine, but also of his own body. How might one obtain maximum energy or intellectual output without overworking the engine? As more and more academic philosophers became acquainted with problems in engineering, they realized the importance of regulating the power running through an engine such as the body. As usual, Babbage made the analogy explicit: right after discussing governors in his *Economy of Machinery and Manufactures,* he noted that the fatigue brought on human muscles by work "does not altogether depend on the actual force employed in each effort, but partly on the frequency with which it is exerted."[48] Natural philosophers hoped to have as much control over their bodies as they did over their machines. They established their credentials as adept body-machine governors—Ure's scientific managers—during the late eighteenth and early nineteenth centuries.

Overworked Vision:
Body Management among Hemionopes

The following four case studies illustrate the increasing importance of steam-engine physiology for understanding hemiopsy and the natural philosopher's body more generally. In many ways, William Hyde Wollaston and David Brewster continued to draw from older bodily discourses based on the balance. We can, however, see in their work the first signs of recognition that using the body always involved problems of inefficiency. The key, more fully realized by George Airy and John Herschel, was to minimize this inefficiency by regulating the levels and uses of the body's power.

I should note here that the experiences of these men were not immediately perceived as instances of the same phenomenon of hemiopsy. The term seems to have originated with George Airy's 1865 paper on his own case.[49] His son, Hubert, gave further definition to hemiopsy and showed that it applied to the previously disparate cases of his father and other natural philosophers in the nineteenth century. I therefore use "hemiopsy" anachronistically in the following sections. This is not meant, though, to mask the fact that before Hubert Airy's synthesis, the symptoms caused considerable confusion. Their cause was not entirely clear, and hemionopes were often unaware for some time that they shared their affliction with others. They began to find common ground in the early nineteenth century through the discourse of steam-engine physiology.

An Anatomical Explanation: William Hyde Wollaston

Stuffed unceremoniously into chemist-physician William Hyde Wollaston's diary from his European tours, a small slip of paper bears a quick sketch of a zigzagged vision he had had and some cryptic notes, including the remark that he had seen this object before (see Figure 4.2).[50] The simple sketch would prove to be a typical rendering of the phenomena produced in a hemionopic attack. In an 1824 paper, he described one of these attacks:

> I suddenly found that I could see but half the face of a man whom I met, and it was the same with every object I looked at. In attempting to read the name JOHNSON over the door, I saw only SON, the commencement of the name being wholly obliterated from my view. In this instance, the loss of sight was towards my left, and was the same, whether I looked with my right eye or my left. The

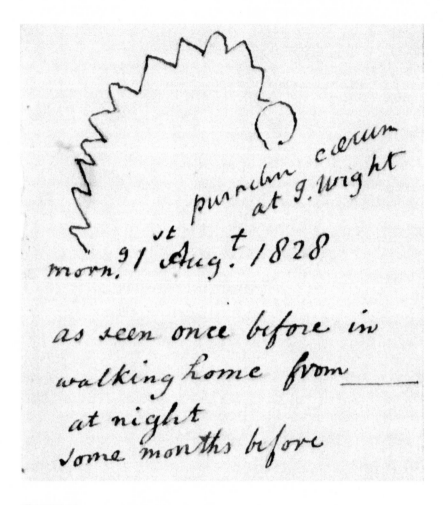

FIGURE 4.2
Wollaston's Sketch of His Hemionopic Attack
On a scrap of paper, Wollaston captured some of the visual experience of a hemionopic attack. The circle signifies the "punctum caecum," or blind spot, which shifted and grew along with the fortification pattern (jagged line) extending out to the left. Source of illustration: Notebook 5, William Hyde Wollaston papers, Ms. Add. 7736, Cambridge University Library. Reproduced by permission of the Syndics of Cambridge University Library.

blindness was not so complete as to amount to absolute blackness, but was a shaded darkness, without definite outline. The complaint lasted only about a quarter of an hour.[51]

Experienced in medicine and optics, Wollaston found himself unusually qualified to remark on the phenomenon and its causes. He borrowed his explanation from Newton's hypothesis that the right- and lefthand side of the optic nerve were each "semi-decussated" or halved, such that each eye connected to both sides of the optic nerve (see Figure 4.3). Hemiopsy, therefore, might be the temporary incapacity of the fibers from one side of the optic nerve. The most plausible cause of this condition seemed to be fatigue.[52] He subsequently invented the highly popular periscopic spectacles in order to save the eyes from the fatigue caused by their regular motion.[53]

After he publicly announced his visual abnormality in 1824, Wollaston's condition worsened. Four years later he noticed visual deterioration in his left pupil and numbness in his left arm. A doctor told him he probably had a fatal brain tumor. Wollaston died a few months later.[54] In their autopsy, Benjamin Brodie and two junior colleagues (George Babington and James Somerville) discovered an egg-sized tumor in the right optic thalamus of his brain, and noted the strange brown color and softness of his right optic nerve. These symptoms were commonly connected with visual malfunction, and in Wollaston they also paralyzed half of his body during the last few years of his life.

Brodie later reflected on the autopsy results as poignant evidence of Wollaston's mental superiority. The tumor, Brodie wrote, had grown so large by the time of Wollaston's death that it might have first taken root in his early years. Somehow, though, the man's career had proceeded brilliantly. On his death bed, though he lay seemingly insensible, he scribbled out some final figures relative to his last scientific investigations.[55]

Ironically, the autopsy posed problems for Wollaston's semi-decussation theory. If his visual episodes had an anatomical cause, it seemed more likely to reside in Wollaston's brain tumor rather than his optic nerves. And anyway, several contemporaries had already pointed out the anatomical impossibility of semi-decussation. However, Wollaston had not restricted his speculations to anatomical structure. He also had hinted that exertion might have as much to do with the onset of hemiopsy as faulty anatomy did. Expanding on Robert Darwin's notion that ocular spectra beset fatigued retinas, he had suggested that his overexertion had stopped flow in part of his optic nerve. Medical school had taught him that overworking the body or the mind could weaken

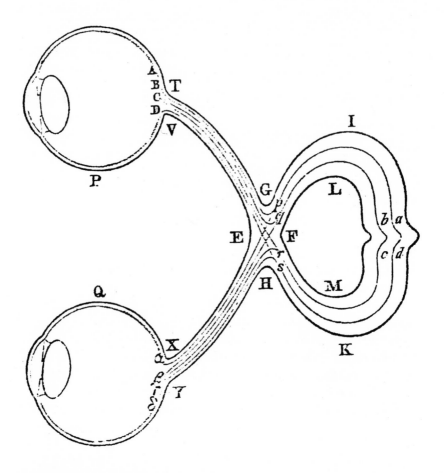

FIGURE 4.3
Semi-decussation of the Optic Nerve
Newton speculated that this schematic captured how the optic nerve connected to the two eyes and produced binocular vision. The "pipes" on the right side of the diagram come from the brain, and split into the right-hand nerve IL and the left MK, meeting again at GF. The nerve tubes from IL split to join the right sides of the two eyes (T and X), while the tubes from the left side go to V and Y in the eyes. Wollaston used this schema to explain how one half of his vision could be obscured, regardless of whether he had one or both eyes open. He suggested that the left half of the optic nerve had failed temporarily. Source of illustration: Brewster, *Memoirs of the Life, Writings, and Discoveries of Sir Isaac Newton.*

them and render them susceptible to disease: "[A]fter violent fatigue of body the mind is not so readily applied to abstruse reasoning—nor is the body after the mind has been fatigued, so capable of its exertions." The best cures were medicines, proper diet, ample sleep, even spas—but above all exercise, of a degree and kind that relieved the patient's mental anxiety and reinvigorated the body. Even after leaving a lucrative medical practice on the Strand in 1800, Wollaston had remained interested in the body and mind's responses to exercise: in 1806 he had performed several experiments on the amount of air he breathed during rest and exertion.[56]

In each of these investigations, Wollaston had tried to optimize his efficiency by approaching his utmost capacity, and carefully studying the limits at which his body ceased to function properly. Wollaston's treatment of the body shows that he had begun to think dynamically rather than kinematically. He had not quite abandoned the notion of the body as a balance, but his interest in the body's external exertions marked a partial turn toward a new way of thinking about the body.[57]

In fact, after his death Wollaston became a veritable poster boy for that rare talent of mixing a methodical approach to scientific investigation with an inspired sense of judgment. Like John Dalton, Wollaston became an exemplar in early-nineteenth-century debates about natural philosophy's future. An older discourse construed Wollaston as a superior body in balance while a newer, reformist discourse adopted Wollaston as a model of skill and mental dynamism who could surmount physical imperfection. For example, in a biographical fragment about Wollaston, MP and philosophical radical Henry Warburton admired his friend's independence from his family's distinguished intellectual and financial history, which would have pushed him toward an "easy" career in medicine instead of a "difficult" one in the natural sciences. "Great . . . were the temptations of indolence to any young man, who like Wollaston was sent to Cambridge, to contend, not with the many for the prize due to superior ability and exertion, but to receive without competition, from the accident of his place of Birth, after a three years residence, a certain income, which though small was enough, should he reside at his college to secure him from want." Rather than allowing his mind to slip into a kind of intellectual gout, the young man soon sought a harder life in chemistry. Thus, he more clearly fulfilled his duty to society and encouraged his physical and mental health to boot. Wollaston, Warburton implied, stood as a privileged counterpart to the self-made men of Samuel Smiles's biographies. In keeping with this image of self-help, Wollaston chose these words to commemorate his exit from the medical profession: "La mediocrité des desirs est ma fortune; l'independence de tout, excepté des de-voirs; est mon ambition" (The modesty of my desires is my fortune; freedom from everything except duty is my ambition [translation mine]).[58]

Not only Wollaston's independence, but also his skill were legendary: his contemporaries jokingly referred to him as "the pope" because he "never proved fallible in any of his numerous experiments, or of his subtile theoretical specu-lations."[59] As such, Wollaston was viewed as the antithesis to another great chemist of his day, Humphry Davy. Davy had defeated Wollaston in the con-troversial election for president of the Royal Society in 1820. On this early bat-tleground between the aristocratic old guard (represented by Davy) and the mixed mathematicians, naval officers, and other practical-minded natural phi-losophers (represented by Wollaston), their two characters came to symbolize a larger rift in the scientific community. In William Henry's 1829 popular chem-istry textbook, Davy epitomized the "bold, ardent, and enthusiastic" romantic with the "fertile and inventive" imagination and affinity for "conjectural analo-gies," while Wollaston rigorously applied his "bodily senses of extraordinary acuteness and accuracy," "vigour of understanding," and "powerful command over his attention" to his experiments. Where one was bold and aristocratic, the other was cautious and diligent (despite his privileged upbringing). Davy con-jectured; Wollaston observed.[60]

Babbage borrowed this exercise in contrasting philosophical character for the conclusion to his *Reflections on the Decline of Science in England*. For the most part, he concurred with Henry's depiction of Davy's and Wollaston's dif-ferences. However, he could not agree that Wollaston's experimental skill stemmed from some innate sensory superiority. Wollaston's "perfect attention" and "minute precision" alone accounted for his uncanny ability to perceive the smallest of chemical particles. If he sported greater vision than most, it was only "one of the natural results of the admirable precision of his knowledge." In other words, the mind—not the body—was the ultimate source of genius, even of the experimental kind.[61]

Henry was stumped by Babbage's refusal to acknowledge the "perfection of [Wollaston's] bodily senses." Certainly, the genius superadded to perception the higher faculties of judgment, memory, and attention. But how, under any acceptable system of inductivism, could one ignore the contributions of the senses to the character of a genius?

> I do not contend [Henry wrote] that in a man of genius the mere visible structure
> of the organs of sense is superior to that of other men—that the eye, for instance,

is a better optical instrument; but I believe that, in the expansion of the optic nerve, in the nervous cord by which it communicates with the brain, or in the sentient portion of the brain itself, there is a superiority of structure, not palpable to the most refined anatomy, which fits it to receive the most delicate and correct impressions from external objects. Sensations thus distinct, when vividly recalled to the memory, become fit elements of new creations. It is the imagination that combines . . . scientific as well as of poetic invention, but no materials can be fit for combination, that have not originally been presented by the senses to the mind, in distinct forms and glaring colours.[62]

Henry deemed Wollaston's gift to be rooted at least partly in his anatomy. This assessment was unusual for nineteenth-century parlance: most natural philosophers by then considered scientific genius either acquired or inherited but housed in nonphysiological faculties such as reason or judgment.[63] Henry's static sense of Wollaston's structural superiority stood in marked contrast to Babbage's dynamic vision of Wollaston's mind governing the body's cues.

Following Babbage's lead, Faraday related the following personal anecdote about Wollaston:

When I was young, I received from one well able to aid a learner in his endeavours towards self-improvement, a curious lesson in the mode of estimating the amount of belief one might be induced to attach to our conclusions. The person was Dr. Wollaston, who, upon a given point, was induced to offer me a wager of two to one on the affirmative. I rather impertinently quoted Butler's well-known lines about the kind of persons who use wagers for argument ["quoth she, 'I've heard old cunning stagers, / Say fools for arguments use wagers'"], and he gently explained to me, that he considered such a wager not as a thoughtless thing, but as an expression of the amount of belief in the mind of the person offering it; combining the curious application of the wager, as a *meter,* with the necessity that ever existed of drawing conclusions, not absolute, but proportionate to the evidence.[64]

Faraday used Wollaston here as an object lesson in the importance of judgment in natural philosophy. The scientific mind did not operate in simple mechanical fashion but actively constructed analogies and hypotheses and made probabilistic predictions when required. That active quality of mind also ensured better observation, since unfiltered sensations were usually unreliable. By implication, Wollaston only strengthened his reputation by studying his own visual defect.

These later sketches of Wollaston's character emphasized the vital place of conscious judgment in the maintenance of health—health of the body and of natural philosophy itself. Static, anatomical structure could answer some but not all questions about the nature of normal and abnormal function. To become managers in the industrial age, natural philosophers needed to demonstrate their dynamic control over malfunction, not to clamor unrealistically after perfect function.

The One-Man Workshop: David Brewster

Himself afflicted with hemiopsy, David Brewster had the benefit of several more decades of anatomical research, which had eliminated the possibility of semi-decussation.[65] In addition, as editor of the *Philosophical Magazine*, he received and published several letters that suggested that overwork might cause such visual phenomena. Henry Kater reported to the journal that a six-month bout with "nervous headaches" had left him blind in his left eye. N. S. Heineken wrote, "Upon rising one morning I observed in the right eye, as it were innumerable faint scintillations or lucid points, the whole field of vision being covered by them. Upon going to a looking-glass, I found that I could not see one half of the face with that eye—it appeared perfectly dark. The effect lasted perhaps half or three quarters of an hour." Physician and natural philosopher William Kitchiner, whose interests lay in both the culinary roots of health and the optics of telescopes and eyes, had told Heineken about seeing such scintillations after over-exerting himself at the telescope. Heineken himself could recall no connection between his own attack and overwork.[66]

When Brewster himself experienced visions, he turned to physiological rather than anatomical explanation. He reasoned that since his images confined themselves to the part of the retina filled with blood vessels, an excess of pressure in those vessels might produce hemiopsy. The following incident placed this hypothesis "beyond doubt" for Brewster:

> When I had a rather severe attack, which never took place unless I had been reading for a long time the small print of the Times newspaper, and which was never accompanied either with headache or gastric irritation, I went accidentally into a dark room, when I was surprised to observe that all the parts of the retina which were affected were slightly luminous, an effect invariably produced by pressure upon that membrane. If these views be correct, hemiopsy cannot be

regarded as a case of amaurosis [blindness], or in any way connected, as has been supposed, with cerebral disturbance.[67]

Neither supernatural forces nor anatomical pathology, then, could be blamed for these visions. Rather, the mundane strain of reading the *Times's* fine print induced them.

In his 1832 popular work, *Letters on Natural Magic,* Brewster devoted two chapters to visual apparitions, demonstrating that they could be given rational explanations. (See the next chapter.) In many cases, however, Brewster had to admit that this was true only in principle; for many of the specific cases, he could not produce an actual, immediate explanation.[68] His vehement rejection of supernatural causes may have derived in part from his own physical vulnerability. To his contemporaries, he seemed somewhat nervous and prone to ill health, probably worsened by the long hours he spent writing the scientific articles from which he earned a living. In a letter to John Herschel, whose privilege he resented, Brewster wrote that his own unstable financial situation required him to conduct his work at a fitful, exhausting pace.

> You who are fortunately free from all the anxieties and distractions of professional pursuits can scarcely form an idea of their disastrous influence on scientific enquiries. While it is your lot to push your researches in calm and uninterrupted seclusion, it is mine to work by fits and starts, now lecturing with breathless exertion, and then abandoning for whole months the ardour of pursuit. You will not wonder therefore if a piece of Machinery so desultory in its movements, and so overloaded at the working point, should perform lazily some of its remoter functions, or even throw itself out of gear from its secondary and often its most agreeable movements.[69]

If Brewster had any particular weakness, remarked chemist J. H. Gladstone of his former mentor, it was that he trusted his mental powers too much.[70] What Gladstone perhaps did not appreciate was that Brewster had never been in a situation to count on someone else's eyes and ears. If Astronomer Royal George Airy represented the perfect example of the scientific factory manager, Brewster had always lived by piecework. He could not afford the luxury of reducing his workload just to keep his "Machinery" from becoming "desultory in its movements."

Life as a one-man experimental shop clearly had its physical consequences. We have already encountered Brewster's ill-fated chemical experiments in chapter 1. In addition, having used his left eye almost exclusively in his optical experiments over fifty years, he had decreased its sensitivity to some lighter shades of red and blue.[71] When he replicated some experiments with the image the sun leaves on the retina (previously performed by Newton and Aepinus), Brewster nearly destroyed his vision. After the experiments, his eyes were "reduced to such a state of extreme debility, that they were unfit for any further trials. A spectrum of darkish hue floated before the left eye for many hours, which was succeeded by the most excruciating pains, shooting through every part of the head." For at least the next few years, he continued to feel the "excruciating pains" in his left eye and head, and had inflammation and reduced "sensibility" in both eyes.[72]

It is illuminating to contrast Brewster's reaction to this vision loss to that of German physiologist of vision Gustav Fechner, who at about the same time severely damaged his eyesight, also while performing afterimage experiments with the sun. Fechner's partial blindness preceded a long and serious neurotic illness. In 1839, he resigned his post as physics chair at the University of Leipzig. In 1843, after three years of living with his eyes wrapped in bandages, he finally found his vision restored and claimed to have gained the ability to see flowers' souls.[73]

In contrast, Brewster acknowledged that his damaged eyesight disqualified him from directly experimenting with many kinds of normal subjective phenomena, but he reserved the authority to reason about them. For instance, he suggested a connection between his condition and the optical illusions brought on by dyspepsia, thus drawing a conventional link between different kinds of nervous disorder. He reasoned that stomach ailments, like his own experiments, exerted unusual pressure on the blood vessels in the eye.[74] Brewster's attitude that he could continue to do good science even while his body failed him (what Gladstone called trusting his mind too much), was consistent with the traditional empirical distinction between sensation and other mental faculties. Dugald Stewart had even elaborated the significance of this distinction for cases of physical abnormality:

> I look thro' a Telescope, and see the satellites of Jupiter, or the Georgium Sidus. I lay down the Telescope, and I can see them no more. Yet still I have the capacity of receiving the visible impression of these objects, and if I again apply the Telescope I am sure of doing so. Here the faculty continues to exist, but is dependent on the organ of perception (the Telescope) to assist it to act. In like manner, altho' the faculties of the mind may be hindered from acting by the disease of the bodily organs, still however these faculties exist, wholly independent on these organs, and may continue to exist after the dissolution of the latter.[75]

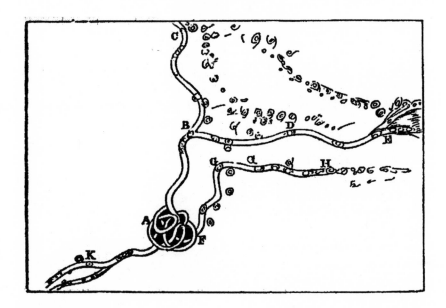

FIGURE 4.4
Brewster's *muscae volitantes*
In October 1838, Brewster made this sketch in order to capture one phase of the changeable visual experience he called *muscae volitantes*. Source of illustration: Brewster, "On the Optical Phenomena."

Stewart used this argument to dismiss the fear that connecting the mind and body meant giving in to materialism, and therefore giving up the intellect's power and the soul's immortality. More immediately for Brewster, the argument helped him maintain cognitive authority. Even if the body-engine broke down, one could still rely on the mind-governor to make sense of and repair the damage.

If the body was often deemed a mere mechanism, the mind never was, as evidenced by Brewster's investigations of what he called *muscae volitantes*, or filaments that he sometimes saw floating in front of his eyes (see Figure 4.4). These phenomena resembled other retinal illusions stimulated by light, and could be induced by looking through a lens with a minute aperture. Brewster positioned a piece of white paper in his line of sight and sketched the images that were "projected" onto the paper, in a process very similar to that employed with a camera lucida. Though others had written on the subject, he argued that they had done so with extraordinary inaccuracy. He promised, through his superior powers of attentive observation and his experience studying the diffraction of light, to give "a correct description, and a satisfactory explanation, of the general phenomena," not just of his own particular experiences with them. By wedging his eye into various positions and reasoning from analogous phenomena in physical optics, he determined that *muscae volitantes* were shadows left on the retina by light diffracted by transparent filaments in the vitreous humor of the eye, contrary to the previous theories proposed by de La Hire, Porterfield, and Mackenzie.[76] To Brewster, knowledge and attentiveness as perfectable attributes mattered more to his scientific authority than the state of his body. He applied the same skills of manipulation and interpretation that he used with all his instruments, and one gets the sense that he conceived of achromatism, hemionopic visions, and a wedge driven next to the eye as equivalent studies in abnormality, which—guided by a superior mind—might say something about normality that the study of normality might never reveal. Thus, while Brewster's limited finances meant he could never be the ideal scientific manager in an age of industrial knowledge production, he believed strongly that he could use the same managerial capacities of judgment and prior knowledge to manage production on the cottage scale.

The Factory Manager: George Biddell Airy

In George Airy we find another case of a natural philosopher easing into a peace with his ever-changing bodily state, though, as we will see, Airy had a whole factory of observers at his disposal that men like Brewster could only envy. To his son, Wilfrid, he seemed the very picture of physical triumph:

> At no period of his life does he seem to have taken the least interest in athletic sports or competitions, but he was a very active pedestrian and could endure a great deal of fatigue. He was by no means wanting in physical courage, and on various occasions, especially in boating expeditions, he ran considerable risks. In debate and controversy he had great self-reliance, and was absolutely fearless. His eyesight was peculiar, and required correction by spectacles the lenses of which were ground to peculiar curves according to formulae which he himself investigated: with these spectacles he saw extremely well, and he commonly carried three pairs, adapted to different distances. . . . In his later years he became somewhat deaf, but not to the extent of serious personal inconvenience.[77]

Even more than his ingenuity, Wilfrid admired his father's ordered life as the key to his success. The elder Airy maintained a regular routine, one that was vigorous but peppered liberally with relaxation: he worked from 9 to 2:30, took a brisk walk, ate dinner at 3:30, napped for an hour, took tea, worked from 7 to 10 in the family room, and finally spent an hour reading recreationally or playing cards before going to bed at 11. To break up this routine, Airy vacationed for about two months each year and always made physical activity a predominant theme of his excursions. Many of his habits seem to have sprung from his years as a sizar at Cambridge, where he took four- or five-mile walks and zealously hand-copied classical literature and mathematical problems onto volumes of paper.[78] In reporting these routines at length, Wilfrid Airy satisfied a contemporary interest in the daily lives of great men, whose discipline might serve as an example to others.[79]

George Airy took this discipline with him to the Royal Greenwich Observatory, when he became Astronomer Royal in 1835. When he first took the post, Airy argued that the recent inefficiency at the observatory had resulted from his predecessor John Pond's ill health, the incompetence of Pond's first assistant, and the consuming task of supervising chronometers.[80] All three factors, he argued, drained energy from an otherwise efficient observatory. Any sign of inefficient operation under his own watch would be corrected immediately. He carefully—one might say obsessively—recorded in his official journals each instance of illness in himself and his assistants.[81] When one of his magnetic observers, Hugh Breen, began to perform poorly on personal equation tests (which measured the deviance of his observations from the norm), Airy instructed an assistant, James Glaisher, to test Breen's vision further. The offended Breen initially refused, thinking Glaisher had capriciously singled him out, but submitted when Airy threatened to fire him. In Airy's eyes, Breen's offense seems to have been his lack of self-awareness more than defective vision. A "personal equation" could be corrected or corrected for—but only if the observer had the presence of mind to notice the discrepancy in the first place.[82]

Airy's persistence in this matter signaled his own equation of the health of the observatory and the observers. He believed he could best manage the health of his enterprise by modeling the observatory on the factory. He divided and routinized much of the labor among various observers, computers, and assistants. This did not mean, however, that Airy "de-skilled" all of the observatory's functions. Airy's own expertise, for instance, enabled the construction of a new transit circle. Likewise, his assistants, generally Cambridge graduates, handled complicated managerial and scientific tasks.[83] Many of Airy's contemporaries admired his managerial prowess, which enabled him to mechanize the observatory's mindless duties, allowing him and his better educated assistants to pursue more theoretical matters. Friend and admirer Otto Struve gushed,

Each of the last numbers of the Monthly Notices [of the Royal Astronomical Society] contains articles of high interest from your pew. Your activity is really prodigious; to your friends it is [at] the same time an agreeable proof of your strength and good health. For me your scientific activity is besides an object of envy. In fact your position which allows you to devote yourselves so earnestly to scientific pursuits is enviable; particularly in the eyes of one who since three years and more moves continually in the thread-mill [sic] of annoying official concerns.[84]

This letter distinctly echoes the content, if not the emotional tone, of Brewster's lament to Herschel, quoted earlier. For Struve and other admirers, the healthy and hierarchical operations of the Royal Greenwich Observatory indicated an equally healthy director whose privileged circumstances enabled appropriate mental direction of the body of observers and instruments.

This division of labor did not impress everyone. In 1847, Airy faced the uncomfortable duty of explaining to the Admiralty why he himself made virtually none of Greenwich's astronomical observations. His former mentor Sir James South tipped off the Admiralty by reporting that of the 69,204 observations made at Greenwich from 1836-44, Airy had only contributed 164. In his defense, the Astronomer Royal argued that "mere" observation, "the lowest of all employments in an Observatory," could be left to his staff. Airy and his assistants had more than enough complicated tasks to fill their working hours.[85]

Greenwich Observatory historian A. J. Meadows has speculated that Airy underestimated the difficulty of observation because of his own poor eyesight. There is some evidence to suggest this is true (though ironically, he once advised a novice astronomer: "*Above all things use your own eyes*").[86] In 1821, when Airy first attempted an account of his astigmatism, the delicacy of his own vision forced itself onto his attention.[87] Only after he had devised a means to correct for his astigmatic vision did he make public his condition (and, notably, his own explanation). By contrast, Airy waited forty years to publicly discuss another unusual visual phenomenon that he experienced, a periodic "double image" that he originally thought might have resulted from a stomach ailment. When this strange phenomenon recurred a month later and many times thereafter, he sketched it and began a serious investigation.[88] His 1865 letter to the *Philosophical Magazine* seems to have been prompted finally by Brewster's published account of his own hemiopsy.

In this letter, Airy related that he had experienced hemiopsy at least twenty times over the course of his career, and knew two others who had had similar experiences (he did not mention who). "From the information of my friends, and from my own experience," he wrote, "I am able to supply an account of some features of the malady which appear to have escaped the notice of Dr. Wollaston and Sir David Brewster." Airy hesitated to speculate on the cause of his attacks, though he found the connection to "overcrowding of business" plausible.[89] He preferred to describe the actual attacks in more detail than had previous observers:

> I discover the beginning of the attack by a little indistinctness in some object at which I am looking directly; and I believe the locality of this indistinctness upon the retina to be, not the place of entrance of the optic nerve, but the centre of the usual field of vision. Very soon I perceive that the indistinctness is caused by the image being crossed by short lines which change their direction and place. In a little time the disease takes its normal type, and presents successively the appearances shown in the following diagram [see Figure 4.5]. In drawing this, I have supposed that the principal obscuration of objects is apparently on the left side; by reversing the figure, left to right, the appearances will be given which present themselves when the principal obscuration appears to be on the right side. (In my own experience, I believe it is an even chance whether the obscuration is to the right or to the left.) The bounding circle shows roughly the extent to which the eye is sensible of vision more or less vivid. Only one arch is seen at one time: the arch is small at first, and gradually increases in dimensions.
>
> The zigzags nearly resemble those in the ornaments of a Norman arch, but are somewhat sharper. Those near the letters B, C, D, E are much deeper than those near b, c, d, e. The zigzags do not change their relative arrangement during the duration of the arch, but they tremble strongly: the trembling near B, C, D, E is much greater than that near b, c, d, e. There is a slight appearance of scarlet colour on one edge, the external edge, I believe, of the zigzags. As the arch enlarges, vision becomes distinct in the centre of the field. The strongly-trembling extremity of the arch rises at the same time that it passes to the left, and finally passes from the visible field, and the whole phenomenon disappears.[90]

Unlike his two unnamed friends, Airy never experienced a headache after his hemionopic attacks. He did, however, experience a temporary inability to speak—what we would now call aphasia. His experiences were common and detailed enough to inspire an excited response from Swiss general and cartographer Guillaume-Henri Dufour, who had read Airy's paper translated in *Le Monde:* "C'est avec un intérêt particulièrement vif, que j'ai lu cet article: car j'ai été plusieurs fois atteint de l'affection optique que vous décrivez. Jusqu'ici, j'en

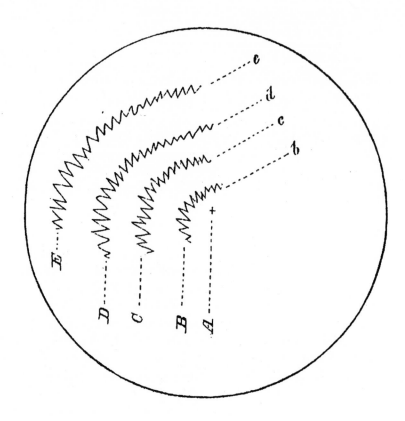

FIGURE 4.5
George Airy's Illustration of One of His Attacks
Like Wollaston's impromptu sketch, Airy's published drawing indicates the zigzag patterns typically seen by hemionopes. Airy added the element of progression from the beginning of the attack (A in the image) to the edges of his field of vision (E). Source of illustration: G. Airy, "On Hemiopsy."

avais parlé à deux médecins qui n'ont pas paru connaître ce cas curieux, et je me figurais qu'il s'agissait d'un accident auquel seul j'étais sujet" (It is with particularly keen interest that I read this article, for I have experienced several times the optical condition which you describe. Previously, I had spoken about it to two doctors who did not appear to know about this curious case, and I imagined that it was an affliction to which only I was prone [translation mine]). The only difference that Dufour noticed was that he had never lost his ability to speak. For relief, he recommended washing the face and forehead with cold water, which had helped him prevent the usual ensuing headache.[91]

Airy's article also prompted a response from Brewster, who found equally interesting the differences between their experiences. Like Dufour, Brewster had never lost his speech during an attack, and in general he thought Airy's case more severe. And though Brewster's zigzag pattern radiated from the foramen centrale, his traveled straight out rather than in a circular arch. He added that he recently noticed another impairment of his vision, which he supposed was related to hemiopsy. He had noticed accidentally that a "considerable portion" of his right retina was blind. This section of his retina sat "about 15° from the foramen [centrale], in a line to the left inclined 45° to the horizon. Its angular magnitude is about 6° in its greatest breadth, which corresponds to a space about a twenty-eighth of an inch on the retina." He did not know its cause or even when he had lost vision there, for the rest of his retina usually compensated for the loss. As before, Brewster expected that his experiments with strong light had caused the blind spot. He concluded that other philosophers and physicians ought similarly to study the changes in their sight and hearing over time, "for it is in the decay or decomposition of organic structures, as well as in their origin and growth, that valuable results may be presented to the physiologist; and facts of this kind have a peculiar value when the patient is himself a practised observer."[92] In this statement, we see an especially bald assertion of the inductive hierarchy that I discussed in chapter 2. We see an acknowledgment of the importance of time and experience to the understanding of the body and its effects on scientific practice. A new pathology posed not just a novel curiosity to the physician, but also a unique opportunity for the natural philosopher to demonstrate his powers of mental management.

Despite the solace from Dufour and Brewster that he was not alone and that his ailment might serve natural philosophy's greater good, Airy's double images particularly plagued him, since they did not submit as readily to his keen powers of instrumental management as his astigmatism had. In fact, the vulnerability expressed in Airy's letter to the *Philosophical Magazine* must have seemed quite striking from someone as confident and accomplished as the Astronomer Royal. Airy feared that hemiopsy was a secondary symptom of a kind of paralysis seated in the brain. He had seen one of its victims left with a mouth distortion which we would now call the aftermath of a stroke. Facing the possibility that some of his most vital faculties as an astronomer — his vision, memory, and speech — might be irreversibly compromised by this disorder, Airy had a few consolations. First, the attacks were temporary, lasting only twenty to thirty minutes, and he had recovered from numerous instances of the phenomenon.[93] Second, his healthy regimen would minimize the effects of defects like his double images. Still, like the other men of science I have discussed, Airy did not always hold up under the strain of his busy schedule. The Astronomer Royal's plate was so full in the mid-1840s that he had to take a convalescent trip to Europe in 1846. Only after this vacation could he turn his attention to John Couch Adams's claims to have discovered a new planet (Neptune).[94] The same year that he wrote to the *Philosophical Magazine* about his hemiopsy, Airy fretted to Norwegian astronomer Christopher Hansteen about a *carte de visite* of himself he had enclosed. He feared the picture would show that "the effects of time upon me are . . . conspicuous," though he rather anxiously consoled himself that physiognomy did not accurately reflect his health: "I feel no decay of strength, bodily or mental."[95] For all of Airy's pecuniary advantages over Brewster, both men had constantly to beat back the nagging uncertainty that their managerial skills were up to the challenge of the demanding industrial environment.

Hemiopsy as the Result of Overwork: John Herschel

The gradual decay of strength over the course of one's life could be frightening, but it could also constitute a heroic sacrifice to the pursuit of truth. When John Herschel began experiencing strange visions in the 1840s, he embraced the latest opportunity to fulfill his physically demanding ideal of scientific work. His astronomically accomplished aunt Caroline had already cautioned him "not to overwork yourself like your dear Father did." Soon after that, Brewster recommended "the medicine that has saved me, — a calm life in summer and a devotion of yourself to rural efforts, or any other engrossing pursuits [that] will keep off the mind from the poison of its own abstractions."[96]

It would seem at first that Herschel failed to heed this advice. In his popular writings, bodily needs frequently took a back seat to the mental appreciation of nature and its divinely designed intricacies. This meant not a denial of the body, per se, so much as its subservience to the reason and will. He deemed this approach important enough to make it the leading idea of his most famous and most popular work, *A Preliminary Discourse on the Study of Natural Philosophy*. Human dominance over nature seemed surprising, he wrote, when one considered only the physical frailty of the human constitution. But it was the exercise of reason, not brute strength, that allowed our dominion. A choice few had reaped from the faculty of reason even greater fruits than the satisfaction of bodily appetites: "Every one who passes his life in tolerable ease and comfort, or rather whose whole time is not anxiously consumed in providing the absolute necessaries of existence, is conscious of wants and cravings in which the senses have no part, ... [and] he will readily admit them to hold a much higher rank, and to deserve much more attention, than the former class." These feelings held a higher rank because they led one to contemplate the perfection of God's creation.[97] This did not mean these faculties entirely transcended the body. Herschel thought of the mind's power of generalization as "a ravenous and hungry mood grasping[,] swallowing[,] digesting and assimilating all that comes near and condensing and pocketing all that is over and above (like Sancho Panca at the wedding feast) for future use."[98] The proper method for understanding God's creation was not to deny the body as an ascetic would, but to discipline it utterly as an industrial manager would.

Though Joseph Agassi rightly called Herschel's philosophy in the *Preliminary Discourse* a "philosophy of success," we should not conclude that he ignored the importance of failure.[99] In a work meant for a popular audience, Herschel understandably presented the natural philosopher in full vigor and perfect freedom from prejudice. In his more specialized and private writing, however, an important part of scientific method was knowing the self, warts and all. He argued in his 1841 review of Whewell's *History of the Inductive Sciences* that disease provided unique opportunities for scientific study; by causing deviations from the bodily norm, it allowed scientists to obtain a clearer understanding of normal processes.[100] Study of one's abnormal body was a precondition for achieving success.

Herschel was able to put this philosophy to good use when his health began to deteriorate. In 1852, when Herschel was at the peak of his career, he forwarded to Airy some observations from his assayer's sister, who had seen some strange optical phenomena out her window one morning. Herschel asked Airy's opinion, adding, "It seems to have been well & carefully observed."[101] Airy's reply is worth quoting at length, in part because it seems to be the only extant account of what the woman saw:

> I [Airy] cannot explain every individual appearance, but I see my way through the general matter perfectly well. The observer [the assayer's sister] evidently had her nervous system thoroughly upset, by cares and want of rest. This is the circumstance which above all others [is] the liability of hemiplegia: my wife, who is frequently attacked by it, recognizes the drawings as characteristic; and some of them (those with the bright spot in one corner) is exactly similar to what she saw a few days ago.

In relation to the object seen at the window, there is one local circumstance to be examined. Does the window front the East? Venus is blazing there intensely bright every morning (and has already brought me some letters—people thing [*sic*] it a meteor. I wish some body would instruct folks in a little astronomy). If the window fronts the East, I conclude that Venus gave the light by which the nervous disturbance of the retina was made visible. Bright sky light would also do it.[102]

Their correspondence on this subject appears to have ended there, for reasons Airy would not know for some time: what Herschel failed to mention in forwarding these observations was that he too had been seeing similar phenomena. Airy himself neglected to mention either his own hemionopic visions, or his fear that he too might have a brain condition like hemiplegia. In fact, the evidence suggests that Herschel and Airy did not openly discuss their own hemiopsy until a full decade later.[103] Rather, for some time the assayer's sister and Airy's wife served as ciphers in this discussion, and of course it is crucial that both these ciphers were women. Male natural philosophers preferred that their own physical frailties have a clear explanation and cures before they became public. Their qualification as managers of industrial-age knowledge production depended on this ability to manage inefficiency in all its forms.

At least as early as 1843, Herschel began periodically to see floating, colorful disks of light which at first he called "ocular spectra" or "fortification patterns" (because of their fort-like shape). The first time he recorded one of these ocular spectra in his diary (1846), it woke him from sleep, and he could find no immediate cause for its appearance. His attacks began in the outside corner of his left eye as a "singular shadowy appearance," which gradually moved to the center of his field of vision. As it came into focus, the shadow took on sharper, zigzagged shapes like Airy's, and often were colored (see Figure 4.6). He could see the shapes whether his eyes were open or shut, and the attacks typically lasted one or two minutes.[104]

Herschel hoped that he could find a corrective—or at least an explanation—for his own optical problem. He must have felt confident (at least at first) that he could do so, since his experiments with the transmission of polarized light through crystals produced colorful patterns that had a appearance similar to the physiological effects Herschel had begun to experience.[105] During his early work on polarization and in astronomy, he also had discovered that he could see objects more clearly if he looked at them indirectly. He reasoned that the center of his retina had become temporarily fatigued from too much use.[106]

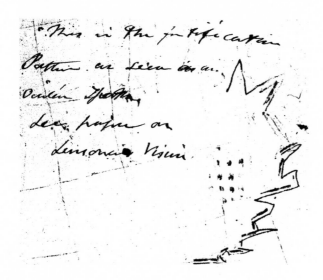

Object—1 cross of the window—clouds moderately bright. Time of exposure 30 seconds

Time hrs.

9.3 .18	All dark
3 .30	bright yellow crossed with dark red
3 .46	very bright yellow
4 .15	Eclipse
4 .25	
35	Eclipse
47	Reappearance crossed yellow
5 .3	Ecli[pse]
5 .11	Crimson crosed [sic] with yellow
5 .23	
30	Rich crimson
32	
53	
9.6 .12	deep Prussian crossed with yellow
	Paler blue crossed with y[ellow]
	beautiful pale blue crossed with greenish yellow
9.8 .47	All disappeared—veil withdrawn
	eyes still closed

FIGURE 4.6
Herschel's Sketch and Notes on "Ocular Spectra"

In hurried, abbreviated handwriting, John Herschel recorded the appearance and development of three ocular spectra in May 1855. (His record of the first of these encounters is reproduced here in the lower half of the illustration.) The first two seem to have been caused by his exposing his eyes to a bright object—probably the sun—for thirty seconds one morning. The third attack appeared that same month while he was in bed. Note that he carefully logged the precise time of each change in the image. Each full attack lasted about four to seven minutes. Table based on material from Herschel, "Ocular Spectra," JH-HRC W0360.

At another time, he sketched what he saw during one of these attacks (top half of illustration). Source of illustration: Herschel, inside front cover of "Photographic Memoranda," 1839-59, John Herschel Papers W0268. Reproduced by permission of the Harry Ransom Humanities Research Center, University of Texas at Austin.

He found more answers through a simple experiment similar to those he designed for color blindness. In a room lit only by a candle, he fixed his eyes on a distant object, then held a strip of white paper vertically about a foot in front of his nose. When he focused his eyes on a distant point, the paper strip appeared double, both images white. If he moved the candle to one side of his face, the strong light in one eye made the corresponding image appear green, while the eye in shadow saw a red strip. The effect intensified the longer he looked. He experienced a similar effect when he worked with colored glasses. He concluded that the candle's stimulation of one part of the retina was strong enough to affect the image of the paper strip on a different part of the retina. He told a friend that "the case seems analogous to those ocular spectra in which a part of the retina having been overexcited by a bright object, appears dark on closing the eyes while the remainder of the retina glows with vivid colours, though it has been sheltered from all excitement."[107]

Over time, then, fatigue—augmented, no doubt, by the heavy doses of laudanum that he took—became the clearest culprit behind Herschel's hemiopsy.[108] He noted in a January 1852 diary entry that he

> had taken 15 drops laudanum at night, 20 night before—had dreamed. Waked—eyes full of superbly rich pattern work [illegible] spectrum of quite a new character—1st large chequers rectangular & separated from each other like several distinct chess boards—with blue
>
> Then conveying [illegible] radiant pattern like feather work with yellows & reds.—several [illegible] changing rapid.[109]

He hastily sketched this experience on a scrap of paper eight months before corresponding with Airy on the matter, and about two years after becoming Master of the Mint. The Mint position and chronic bronchitis contributed to a nervous and physical collapse in the mid-1850s, after which the hemionopic visions increased in frequency. He spent much of his remaining fifteen years in bed or in a wheelchair enduring repeated bouts of bronchitis, rheumatism, and gout. He found himself in a "state of languor & depression both of bodily & mental activity—which deprives me almost of the power of doing anything calling for consecutive thought or giving *anything* its due attention be it ever so pressing."[110] More insistently than before, Herschel begged off public obligations, including his advisory role in selecting faculty for the new University of Melbourne. The task, his daughter Isabella said, hung around her father's neck like a "millstone."[111]

Herschel struck many as a sickly fellow. Already in 1838, upon his return to England from his four-year stay in the Cape Colony, he appeared haggard. Spotting him at the British Association meeting in Newcastle that year, a *Daily News* journalist was surprised by the "deep lines of thought on his face," the "straggling grizzled hair," "absent expression of the eyes," and his "restless manner."[112] On a visit to the Herschel home at Collingwood five years later, novelist Maria Edgeworth worried about her friend's well-being:

> He shewed us a Daguerreotype of the stand of the great instrument [William Herschel's forty-foot telescope] before it was taken down. He told us that the impression of that frame as it stood was so strong on his eyes that when it was gone he some time afterwards saw it in its place so plainly before him that he thought he could have touched it. There it is still! I never saw so sensitive a person — almost too much for his health. He complains — no he never complains, but he told us of a strange delusion or disease of his sight which comes on at night sometimes when he is sitting up reading. The farthest part of the room vanishes and by degrees the circuit of sight diminishes so that he can at last see only the table before him and just the space occupied by the candles. This should warn him not to overwork and it does warn Lady [Margaret] Herschel to take all means to prevent his overstraining his great faculties.

If perception were like photography — a science in which Herschel showed a concerted interest — there was at least the hope of an explanation and the ability to control the situation. Indeed, Herschel spoke of his own visions as the result of a "photographic process."[113]

In a more public effort to explain his visions, he spoke in 1858 to the Leeds Literary and Philosophical Society on "sensorial vision." To this audience, he argued that his visions were the result of an especially active mind, and speculated that the site of his problem was the *sensorium commune*. He mentioned a woman that he knew (presumably the assayer's sister) who had seen similar phenomena; he noted, however, that her visions always followed "a violent headache," something he did not experience.[114] Implied in his talk was the notion that anyone might suffer nervous illness but the philosopher had the resources of reason and experience at his disposal, tools that could diagnose and even cure. For example, he argued that the subject of ocular spectra was "far from being exhausted, and it is to the habit of attention to such sensorial impressions, fostered by frequently watching the development of these spectra *under a variety of circumstances* in my own case, that I attribute my having been led to notice that other class of phaenomena of which I shall presently speak, and which from their inconspicuousness, I suppose, *escape the notice of most people*."[115] In this statement, Herschel both validated his own authority on the subject of vision, and encouraged those with similarly sophisticated skills to inquire further. Privately, he wondered if his visions might indicate "an intelligence working within us which is not our own." After the lecture, Augustus de Morgan's wife, Sophia, a mesmerist, wrote to Herschel of a friend's experience with the fortification patterns of "sensorial vision" and also of Richard Sheepshanks's daughter's visions of faces — more complicated apparitions than normally brought on by hemiopsy.[116] Herschel's investigations thus left a number of questions unanswered. In particular, what, if anything, linked together the various afterimages that so many early-nineteenth-century natural philosophers had experienced? And would these men's intellects really hold up under the strain of industrial production?

Hubert Airy's Synthesis

Like Robert Darwin, Hubert Airy found in his father's visual studies a subject for his medical thesis. In fact, for the Airys, the "morbid phenomena" of hemiopsy were a family affair. As George had noted in the letter to Herschel discussed above, his wife Richarda's hemiplegia often instigated these attacks. Hubert credited his own first attack, on 6 October 1854, to his earnest struggle with some trigonometry just before the Michaelmas holiday. Of its exact cause he knew nothing, except to say that it certainly seemed related to extreme habits of exercise and diet, which somehow affected the brain.[117] During another episode in March 1868, he rushed to the library at Guy's Hospital. Holding a notebook eight inches from his face, he traced the still-present zigzags as they "floated" on the paper.[118] As the pattern changed, he sketched its development (see Figure 4.7). His attacks were singularly unpleasant. At their height, his eyes "seemed to be boiling over with light," and they left him "stupid" with headache afterward. Otherwise, we can conclude "like father, like son": Hubert blamed the March 1868 episode on a five-hour walk he had made that morning on a nearly empty stomach.

Eager to make more sense of these phenomena, the young medical student collected the various published accounts of severe and chronic ocular spectra, and solicited some further observations from Herschel.[119] Herschel, by then quite weak and only three years from death, found the very conversation with Hubert about hemiopsy enough to trigger an attack.[120] As an aspiring

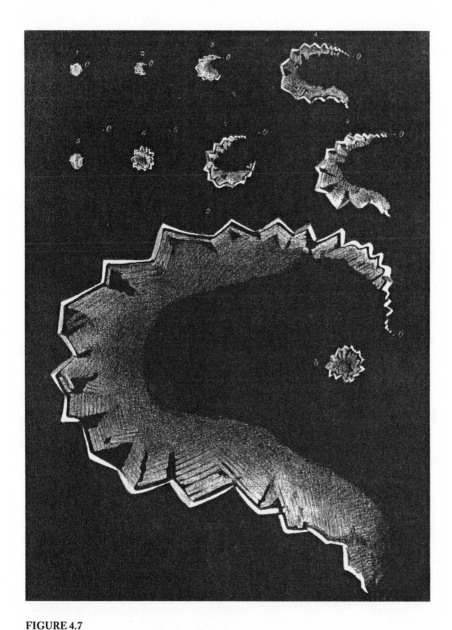

FIGURE 4.7
Hubert Airy's Depiction of Hemiopsy's Progress
Even more vividly than his father or Wollaston, Hubert Airy represented the progressive growth of the fortification pattern seen during a hemionopic episode. The attack began with a small, jagged spot in his vision that obliterated external objects, as at above left. The pattern grew to full size as in the bottom half of the illustration. Source of illustration: H. Airy, "On a Distinct Form."

physician, Airy hoped to find information about the problem in the medical literature as well, but was largely disappointed. Substantial material existed on hemiplegia and its deterioration into aphasia and blindness. But few physicians noted the specific symptoms of hemiopsy,[121] and those that did, did not uniformly classify it. Some physicians mentioned it briefly as a disease of the eye (as in Frederick Tyrrell's *Practical Work on the Diseases of the Eye*), an occasional symptom of "sick-headach" (John Fothergill), or an unclassified phenomenon. Fothergill's analysis appeared in an unpublished paper of Caleb Hillier Parry's, which became available in print in 1825. Parry, a physician at Bath General Hospital in the eighteenth century, frequently fasted for eight to ten hours, and this combined with the fatigue of working sometimes produced "a sudden failure of sight."[122]

Hubert Airy postulated that all these cases belonged to the same malady, which he called *transient hemiopsy*. This term both distinguished the disorder from the related, permanent version so well studied by ophthalmologists, and denoted the disorder's tendency to cloud half the vision. He further divided transient hemiopsia into two closely related types: the first, described by Wollaston, Arago, and Brewster, and the second by his father, himself, Dufour, Herschel, and Charles Wheatstone (on whom more in a moment). His father's 1865 paper, as the first published account of the second kind of hemiopsy, mentioned features never described by Wollaston, Arago, or Brewster. "The outward spread of the cloud, its arched shape, its serrated outline, with smaller teeth at one end than at the other, its remarkable tremor, greater where the teeth are greater, its 'boiling,' its tinge of scarlet, and its sequel of partial aphasia and loss of memory, are all new features."[123] Not entirely satisfied with *hemiopsy* because it suggested the false semi-decussation theory, Airy offered another possible term: *transient teichopsia*, indicating the victim's vision of the zigzag pattern that looked like a fortification or town wall. Like most of his predecessors, Airy lacked a specific physiological theory as to the causes of, or remedies for, these attacks, but suspected a connection with too little or too much exercise, too little sleep, or even windy weather. He also adopted a quasi-phrenological stance when he hinted that a localized portion of the brain was a more likely agent for these attacks than "nervous sympathy" or "gastric derangement," which both seemed too vague to explain such "definite" and "localized" symptoms.[124]

Airy presented his findings to the Royal Society on 17 February 1870,[125] and Wheatstone recommended the paper for publication in the *Philosophical Transactions*. Though Airy had been beaten to the presses by other observers,

Wheatstone thought hemiopsy had "never been so completely made the subject of scientific observation as it has been by the present writer."[126] Wheatstone suspected that hemiopsy was far from uncommon, but that its transience made it extremely difficult to describe with enough precision to warrant publication. Furthermore, those who usually paid little attention to vision or optics were apt to dismiss the attacks as inconsequential. He also took the opportunity to reveal his own experiences. While writing a letter one evening in 1849, Wheatstone recalled, the characters near the center of his vision had become invisible. This spot of invisibility spread to the left in both eyes, and lasted a half hour, after which Wheatstone felt faint. He described having these episodes frequently, usually while reading. Privately, he noted several other instances of these "mental" or "ocular spectra," which sometimes occasioned "rocking of the brain." Since Wheatstone had no intention of publishing anything on the matter himself, he gave Airy permission to use his letter as further evidence for his case. Airy's paper, Wheatstone supposed, would recall similar experiences to the minds of many.[127]

As we have seen, the paper must indeed have inspired recognition in many early industrial natural philosophers. But by 1870, Airy could speak with more confidence of his colleagues' ability to manage such pathology. As we saw in the first chapter, he believed that natural philosophers' "habits of accurate observation" and unusual exposure to the "risk of impairment . . . of the eyesight" imbued them with "especial advantages for the study of visual derangements."[128] Men of science — now more securely entrenched in the state bureaucracy, universities, and popular entertainments — still needed their public to believe in their superior managerial skills. As Airy recognized implicitly, however, they now asserted this authority from positions of institutional strength.

Conclusion

As tales about the changing nature of scientific work in nineteenth-century Britain, the cases of hemiopsy discussed here deepen our historical understanding about the mind-body relationship, the division of intellectual labor, and visions of masculine work in the sciences.

First, I have noted how conceptions of the scientific mind borrowed from those of the engine governor and the factory manager. Simply because the understanding of the body as a machine persisted throughout the modern period, does not mean the nature of that comparison remained static. Whereas

FIGURE 4.8
Ampère Chasing His Equations
A story about the French *physicien* André-Marie Ampère amused those who thought of natural philosophers as absentminded. While walking one day, Ampère thought of a way to write out his formula for the electrical unit of measurement. He stopped to sketch his ideas in chalk, only to realize as his equations pulled away that he had been writing on a carriage. Ampère eventually caught up with the "runaway formula." Source of illustration: Hawksworth, *Workshop of the Mind.*

the Enlightenment mechanism moved automatically as long as its parts were well constructed and balanced, the nineteenth-century steam engine was more than the sum of its parts. The engine did not run passively, but actively changed its environment through the governor's direction. Power and change, not balance, were its key features. The mind served as a necessary governor over the body-engine's change in time. The scholar sometimes became so absorbed in his work that he overloaded the machinery. This "absence of mind" was generally harmless, so long as it did not interfere with production. The then "well-known" story of Robert Hamilton, a professor at Aberdeen, revealed antics such as his not recognizing his wife in the street, wearing mismatched socks, and apologizing to a cow after running into it. The moral: a wise man would train his mind to notice such overextension and relieve it through play, exercise, and travel (see Figure 4.8).[129]

Between the Enlightenment and the mid-nineteenth century, all but the most radical British materialists and spiritualists retained their faith in dualism. But the mind's governing *role* over the body did change in the early nineteenth century. The British natural philosopher's view of the mind changed from that of a political ruler to that of an industrial overseer. If the ideal function of the former was to maintain harmony, the latter was supposed to regulate power with the maximum possible efficiency. The system remained hierarchical. Fact-gatherers (analogous to the automatic mechanism and the factory laborer) formed the backbone of research, but only the elite philosopher could direct the growth of knowledge. In dismissing phrenologist George Combe's loftier metaphysical claims about the mind, Brewster wrote that those "who make the parts of a machine [e.g., fact-gatherers such as Combe] are not generally those who can construct the whole."[130] Likewise, Herschel, Airy, Babbage, and others cast a Smithian eye on the labor force for natural philosophy. Legions of amateurs would serve as Baconian fact collectors, and those facts would then be elevated into theoretical knowledge—that is, given meaning—by an elite cadre of mathematicians, experimenters, and theoreticians.[131] Analogous to the mind governing the body, and the manager governing the factory workers, the philosophical elite would extend the higher faculties of reason and educated judgment to the data recorded by their army of observers.[132] Rather than passively watching nature's operations, the dynamic philosopher hypothesized, poured, adjusted, inquired, and above all synthesized. The moves by early-nineteenth-century natural philosophers to eschew judgment from the natural sciences were meant to apply only at the ground level of fact collecting.

One of the most powerful aspects of this reconception of nature's physical forces and the scientist's body was that it emphasized activity and power over time (physiology) instead of static structure (anatomy). Whewell in his *History of the Inductive Sciences* denigrated past "stationary" periods in science for wishing to know the whole of nature "at a single glance" instead of relishing the process of developing their knowledge. Humans and the building blocks of matter were alike in their potential energy, which in interaction with nature (e.g., experimentation) became dynamic. Consider, for example, John Tyndall's similar descriptions of the mountaineer: "What would man be without Nature? A mere capacity, if such a thing be conceivable alone; potential, but not dynamic," and then his account of gravity acting on a weight: "While in the act of falling, the energy of the weight is active. It may be called actual energy, in antithesis to possible[;] or dynamic energy, in antithesis to potential."[133] Man's and nature's dynamism simultaneously produced knowledge, power, and health.

So, the body was cast as a machine, but one that wasn't reducible to the mechanical models of the seventeenth and eighteenth centuries. Just as the modeling of microscopic forces had taken a phenomenological cast, so too did the understanding of the nervous power of the body and mind; both conceptions retained a certain quality of ineffability, near indescribability. The ineffability of experience and physical modeling was considered uneradicable, even formative, of physical knowledge. "About the idea of Force, I hardly know what to say," Airy wrote Herschel in the 1860s, "It seems to me that in general our ideas on the metaphysical origins of such things (mine at least) are shaped by our physical experience, and I certainly feel this in regard to the action (of the nature of attraction) of one body on another."[134] The importance of experiencing an experiment, of taking part in its energy, might contribute to the uniqueness of experimental culture, in which the work "relies on a kind of subtle judgment that is notoriously ill-suited for the prose of hypothesis and deduction. . . . Only the experimentalist knows the real strengths and weaknesses of any particular orchestration of machines, materials, collaborators, interpreters, and judgments."[135]

While British natural philosophers generally shared this ideal of the energetic philosopher, they did not necessarily agree on how to realize that ideal. In their scientific habits, Wollaston, Brewster, Herschel, and Airy embodied very different approaches to balancing the healthful physical activity with the authoritative, elite practice of contemplation. With relatively little assistance, Wollaston and Brewster hammered away at experimental problems, diving headfirst into the physicality of natural philosophy. In much of his work—for instance, in photographic developing—Herschel also followed this workshop model. But in other areas of his research he preferred to sit at the head of a factory-like enterprise—for instance, in his development of physical observatories. Airy's presence was even more managerial, less hands-on, though he repeatedly called attention to the toll that astronomy took on him. Airy configured his managerial, mental work as physically exhausting. But he denigrated what he saw as lowbrow excessive physical involvement in experiments. After seeing Brewster's *Treatise on Optics* and the *Edinburgh Journal* he edited, Airy remarked that his "estimation of Brewster has sensibly dropped. What an insolent fellow he is, on the strength of experiments which are very valuable but which he does not know how to interpret (for nothing can be more awkward than his theories) to set about abusing the rest of the world in that way. There are many persons at Cambridge who understand the subject much better than he does, though they have made few experiments. . . . Really these gentlemen of the northern schools ought to be taught better."[136] To Airy, direct experience did not authority make. If that were true, any one of his assistants at the Greenwich Observatory had grounds to usurp his position. Airy's mathematical education preserved him from such a fall from grace, and underpinned his sneer at Brewster's workshop mentality. Despite these important differences of scale, however, Brewster and Airy and the rest each sought ways to incorporate their unusual bodies into the industrial apparatus of natural philosophy.

Finally, we can use this analysis of changes in mind-body dualism to reconsider masculinity in natural philosophy. In one sense, masculinity did not change from the Enlightenment to the mid-nineteenth century, for in both periods male physiology and psychology remained the norm. But since for men of science the normal mind-body relationship had changed from a balancing act to a ponderomotive process, so did the masculine image of the natural philosopher. The eighteenth-century man's very maleness allowed him to balance, for instance, refined sensibility (which was accessible to women) and public activity (which generally was not).[137] But the nineteenth-century philosopher had to concern himself more with the management of energy expenditure.

Brewster posthumously cast François Arago as one of these new men. In 1820, the French government hired Arago and Pierre-Louis Dulong to improve the construction and safety of steam boilers. "This task," Brewster wrote, "which was executed with much ability, was as dangerous as it was difficult. The bursting of boilers to which they were constantly exposed, and that too in a limited locality, was more hazardous than that of shells in a field of battle; and while military officers who assisted them — men of tried courage — grew pale and fled from the scene, the two savans went on coolly making their calculations, and observing the temperature and pressure with boilers every moment on the point of explosion."[138] Arago symbolized the manly ability of the energetic philosopher to manage physical stress and produce knowledge. Naomi Oreskes may be correct when she argues that heroism more than objectivity has defined the modern, masculine dominance of the sciences, for the nineteenth-century philosopher used his unique coolheadedness to navigate high-pressure situations in the laboratory and within his own body.[139] Masculinity, like science, was now an achievement always in process, not a stable quality.

FIVE

Rational Faith and Hallucination

A lady once asked me if I believed in ghosts and apparitions? I answered with truth and simplicity; *No, Madam! I have seen far too many myself.*
— Samuel Taylor Coleridge, *The Friend*

WHILE SERVING AS LIBRARIAN to the Earl of Shelburne in the 1770s, the radical chemist Joseph Priestley was called to the chambers of the earl's son. The youth had spent a sleepless night due to a distressing dream in which a hearse carried him to his family's burial place. Dr. Priestley assured him that the dream was simply the product of a fever. The medical attendant assigned to oversee the boy's convalescence stopped in one cold January day, only to find his charge outdoors, running to meet him. As the attendant rode toward the boy to admonish him for his carelessness, the young figure suddenly vanished. Once inside the house, the attendant learned to his astonishment that the child had just died—in his bed.[1]

Observers of industrial-age England have agreed that it was a haunted, hallucinatory place. Carlyle called belief in ghosts one of the "signs of the times." When reports emerged in 1762 of a ghost haunting a house on Cock Lane in the East End of London, Andrew Lang groaned that his contemporaries had a greater obsession with spirits than the medievals had had.[2] After about 1790, ghost stories sold widely as cheap literature. The spiritual had featured especially prominently in Gothic and Romantic fiction, but continued to haunt Victorian writing.[3] These tales made such scintillating reading not just for their luridness but also because they literally captured this period's spirit of rapid change. Vanessa Dickerson has aptly described the ghost as an emblem of

Victorians' feelings of transition between "medieval god and modern machine, monarchy and democracy, religion and science, spirituality and materiality, faith and doubt, authority and liberalism."[4]

What Dickerson captures in her understanding of apparitions is their liminality. In this chapter, I will argue that even before the Victorian period, the experience of hallucinations served as a lightning rod for discussions of the religious and political changes occurring in Britain. The rapid growth of non-Anglican religions from about 1780 to 1860 was one of the most marked changes in this period. That proliferation threatened a "universal" (moderate Anglican) understanding of faith in much the same way that provincial power threatened centralized state government or steam explosions threatened industrial efficiency. Natural philosophers and physicians advocated treating hallucinations as psychological and optical phenomena in order to keep more radical, potentially factionalizing religious understandings at bay. They argued that using one's reason — and by extension, adhering to a moderate faith — to understand visions made better political and intellectual sense than succumbing to the superstitious and backward interpretation of visions as communications with the spiritual world (see Figure 5.1).

Hallucination had become a medical issue early in the eighteenth century because just before this, portents had been reclassified as either natural events or miracles, the latter of which had lost much of their evidentiary status. In other words, from this point on, European churches and natural philosophers largely agreed that events could only have natural or divine causes. Preternatural phenomena — events caused by demons, angels, or other supernatural beings — now could only exist in the imaginations of the vulgar.[5] The new experimental philosophy's admission of singular experiences as typical instead of deviant contributed to this trend.[6] Similarly, German Berrios has pointed to hallucinations' "loss of semantic pregnancy" in the eighteenth century, or the loss of the "belief that their *content meant something,* that [they were] portent[s]. Hallucinations thus became but symptoms or markers of disease and what the patient saw or heard had no longer meaning in itself." This strategy achieved several goals: it helped the sciences and medicine avoid participating in factious battles between different religions; it posited a central authority (reason) to counteract the provincializing forces of superstition and sectionalism; and it allowed natural philosophers who knew they were sane to make sense of their own hallucinations.[7] Well into the nineteenth century, though, the medical and scientific establishment felt the need to continue to denounce supernatural explanations for hallucinations. This suggests that, as with the

FIGURE 5.1
Insane Superstition versus Rational Religion
Taking a healthy, rational approach to hallucination did not mean forgoing one's faith. On the contrary, natural philosophers thought that belief in the supernatural and false ideas went hand in hand. Those who fell into such a trap, like the head of the household at left, were not far from the insane asylum, seen out his window. The patriarch at right has learned to keep his hand on the true Word and his eye on the church, thus ensuring that neither he nor his family will stray. Source of illustration: Emmons, *Philosophy of Popular Superstitions.*

other disciplinary projects that I have described in this book, natural philosophy's attempts to tame hallucination and Britain's religious culture posed a continuous challenge throughout the industrial age.

We might wonder why such antisupernatural polemics continued to appear so long after the Scientific Revolution's supposedly resounding defeat of natural magic. I argue here that the industrial age's volatile religious and political atmosphere inspired a renewed fear among natural philosophers and physicians that the masses clamored after the irrational, both in its materialist-Jacobin and evangelical forms. A wave of literature that appeared between about 1780 and 1860 encouraged the rationalizing of apparitions. These writers did not deny the possibility of miracles, but sounded quite deistic in their claims that only very rarely did a phenomenon defy natural law. Thus emerged yet another strategy for bringing subjective knowledge under the more rational governance of natural philosophy.

The religious climate in Britain was changing dramatically. After the scare of the French Revolution came the Church of England's decline in power and a concomitant proliferation of alternative Christian sects. This altered not only the religious landscape but also the political one. In this volatile atmosphere— as during the Restoration—many men of science saw the political prudence in nonsectarianism. Even to relatively tolerant, moderate Anglicans, the prolife- ration of religious sects in Britain in the eighteenth and nineteenth centuries threatened the possibility of a consensus on truth. Sectarianism of any kind proved so antithetical to the pursuit of universal, natural knowledge, that most scientific societies forbade discussion of religion and politics at their meetings.[8]

Most British natural philosophers took a moderate theological and episte- mological stance on hallucinations. Hallucinations occurred frequently enough that they could not meet Anglican standards for miracles, and enough sane people experienced them that older imputations of insanity did not apply.[9] Neither revelation nor insanity, in other words, fit the bill. Thus, in the first half of the nineteenth century, natural philosophers and physicians carved out a space for sane hallucinations. This project drew much of its justification from a set of moderate and liberal positions that God rarely, if ever, intervened in the course of nature. Deists, agnostics, and atheists recognized nothing but natural causes in the universe's everyday operations. But even voluntarists, who wished to maintain the possibility of divine causality in nature, knew that such miracles would be rare; their rarity would help preserve them from absorption into natural law. In fact, since the seventeenth century, it had been common- place to argue that miracles had ceased long ago.[10]

Natural philosophers often proclaimed that only the superstitious still be- lieved in spirits. In real life, they sometimes found it difficult to explain their own eerie visions. Attempts to generalize the subjective experience of halluci- nation encouraged as much rhetorical flourish and private doubts as our previ- ous two cases of color blindness and hemiopsy. The most popular strategy for escaping these doubts was to argue that those who possessed and cultivated their reason could distinguish between a hallucination and a real object. Once hallucinations were understood as lapses of reason, the need for reason as a weapon against idiosyncratic and divisive experience became that much more apparent. Ultimately, natural philosophers hoped to wield reason as a weapon to defend science and theology against the turmoil that wracked industrial-age British politics. In so doing, natural philosophers once again called their own nervous vulnerability into service. By mastering hallucinations and the volatile religious-political climate that produced them, natural philosophers could unify a fragmented nation just as they claimed to do with provincialism through color blindness, industrial economy through mechanical efficiency.

THE RELIGIOUS CONTEXT:
RATIONAL ANTISECTARIANISM

Traditionally, each of Britain's countries operated under only one official church. Jews, Catholics, and dissenting Protestants had all existed in significant numbers during the early modern period, but it was not until the late eighteenth century that minority religions made serious gains in numbers and voice. Before this, Anglicans, particularly of the latitudinarian variety, acknowledged the value of reason as applied to faith, but they frequently used this logic to persecute rather than tolerate non-Anglicans. As we saw in chapter 3, certain practices among dissenting religions seemed so odd to mainstream Anglicans that they were dismissed as backward and superstitious. For example, London surgeon Walter Cooper Dendy argued that like hysterical women, Quakers were prone to "imitative monomania." Their "religious fanaticism" threw proselytes and preachers alike into frenzies, so violent that their subjects seemed possessed by demons.[11] Such excesses were matched only by the Catholics' "Popish superstition." Believing in the immortality of the soul was one thing, but to believe that God allowed souls to walk the earth was downright savage.[12] In contrast, Britain's established churches claimed to prefer "a cult of moderation, between the authoritarianism of Rome and the excessive individualism of radical Protestantism."[13]

Against this traditional view, religious toleration increasingly became the norm among the educated classes during the period 1780–1860. Superstition, it was argued, should be discouraged, and reasoned religion encouraged, in any possible form. A relentlessly idiosyncratic faith, whether enthusiastic or materialistic, could never lead humanity from its postlapsarian ignorance to a fuller grasp of God's knowledge, but a limited toleration would acknowledge what advocates saw as the reasonability of some non-Anglican faiths. The new tolerance took various forms such as opening universities to dissenters, some liberalization toward Catholics, and disestablishing the Church of Ireland.[14] What began as a minority, "enlightened" position, gained steam throughout the early nineteenth century, even outside radical circles. These reformers advocated not a return to "superstitious" and revelatory religious practices but a rational basis for faith that would transcend sectarian boundaries. David

Hume's argument that religion encouraged superstition and could not serve as a proper basis for government was considered radical in the extreme. But even those repulsed by the irreligion of the Enlightenment *philosophes* could agree that sectarianism impeded good government and national unity. Intellectuals as diverse as Edmund Burke and Christopher Wyvill argued that greater toleration of non-Anglicans was in the best interest of the public order.

In the midst of its long war with Revolutionary France and popular unrest due to economic hardship, Britain could ill afford public disorder. The prospect of tolerating dissenters seemed like a far more palatable choice. To many moderate and liberal intellectuals in early-nineteenth-century Britain, the rational understanding of religion seemed to promise the greatest social benefit. Rational faith still implied the eradication of superstition, but that purpose was no longer seen to be served by the persecution of particular sects. For the sake of reason and political safety, holding nonsectarian views in one's public life had a vast appeal.[15] The broad or liberal position, though it certainly had plenty of critics, dominated many of the centers of scientific power, such as Cambridge, Edinburgh, and the British Association. Accordingly, it became common practice in the early industrial period to disallow political or religious discussions so as to avoid sectarianism at the meetings of scientific societies.[16]

Natural theology, with its strong grounding in reason rather than revelation, served as an important interreligious glue adhering natural philosophy against divisive forces. It had played this role in the Restoration as well. For example, John Henry has contended that the arguments for active principles in natural philosophy grew partly out of a desire to denounce both religious skepticism and enthusiasm: "If God was no longer required to supervise the running of the universal machine, the atheist might argue that God had never been required since the universe may have existed eternally." On the other hand, natural philosophers' confidence that experimental techniques could lead to knowledge about active principles "ensured that the supernaturalism of religious enthusiasts could not assert itself as the truest account of the world."[17] Two of the natural philosophers who held this very position—Joseph Glanvill and Henry More—also gave early, rationalist accounts of superstitions as part of their larger attempt to battle atheism and skepticism.[18]

Despite its increasing popularity in the early modern period, this position had its dangers. Taking a nonsectarian approach in natural philosophy then meant discussing God as little as possible. Apart from showing how the study of nature proved its divine design, little could be said about God's relationship to creation without entering into a theological morass. To deny God any role except that of creator, however, appeared very much like deism. And deism seemed but a short step from atheism. We find an excellent illustration of this ambivalence in William Paley's natural theology and the public reception of it. Though an Anglican and social conservative, Paley effectively allowed God only an abbreviated role in worldly affairs. In addition to understanding God through revelation, Paley advocated looking to the utilitarian principle as the will of God and otherwise structuring faith according to reason. Evangelical critics found the direct experience of God in Paley's universe too pale.[19]

If few nineteenth-century natural philosophers were committed deists, their rhetoric could easily give a different impression. Consider another example: early Newtonian natural theology had walked a middle path between deism and theism. By the early industrial era, however, natural philosophers generally sought to ground all natural phenomena in laws that did not require God's continued intervention. This change happened for several reasons: first, gravity came to be accepted as a given rather than needing a mechanical or providential explanation. Second, Lagrange, Laplace, and other Enlightenment-era French natural philosophers worked out anomalies elsewhere in Newton's system that previously required providential explanations. And finally, the clockwork universe entered the popular imagination, especially as the Glorious Revolution of 1688 grew more distant in memory and the possibility of a lawful, self-regulating universe seemed more plausible and desirable.[20] In the early nineteenth century Cambridge-based reformers turned these developments toward scientific and theological reform. The Analytical Society based its activities on the Cambridge chapter of the British and Foreign Bible Society, which distributed the Bible without its usual Anglican-endorsed interpretations and prayer book. They also sought to rid the Scriptures of its more supernatural elements. Hoping to emulate this democratic enterprise, the Analyticals promoted the study of Lagrangian mechanics as a means to take the mystery out of calculus and distribute it to a wider audience.[21]

Even those such as Whewell who bridled at the "analytic revolution" agreed to a rather limited voluntarist theology. Voluntarists held that God's will formed the basis of theology, and that God actively—though not directly— governed the universe as the ultimate cause. Natural laws did not limit God's actions, and so natural philosophy could not explain the miraculous. So, for instance, in remarking on Laplacean astronomy for a British audience, Whewell argued that a solar system that did not need divine tweaking served as much better evidence of design than one that did.[22] Likewise, Thomas Reid had argued that "upon the theatre of nature we see innumerable effects, which require an agent endowed with active power; but the agent is behind the scene. Whether it be Supreme Cause alone, or a subordinate cause or causes; and if subordinate causes be employed by the Almighty, what their nature, their number, and their different offices may be are things hid, for wise reasons without doubt, from the human eye."[23] Natural philosophy did not, and should not, concern itself with final causes. For Boyle, Descartes, Newton, Whewell, Joule, the Thomsons, and other voluntarists, "allowing" God the power of direct intervention in practice usually meant little. In practice, virtually no event occurred outside the lawful, spirit-free grip of natural philosophy. Thus, systems like the clockwork universe, "though lodged in theologies of nature that remained Christian in inspiration, were to appear perfectly at home when lodged in deistic philosophies."[24] God, just like the managerial natural philosopher, ordered his world through the exercise of his will—all the while making his direction seem effortless and natural.

Even many of the critics of overly rationalized religion did not wish to dim the explanatory power of natural philosophy. Humphry Davy, Britain's greatest advocate of Romantic natural philosophy, once dressed himself in a white sheet in order to convince his grandmother of the unreality of ghosts and the gullibility of their believers.[25] The evangelical David Brewster—though he vehemently castigated Whewell for his Kantian epistemology and especially for his rejection of the possibility of life on other worlds—would also argue that supernatural explanations were rarely appropriate in explaining apparitions.

Joseph Priestley's notorious materialism—developed during his employ by Lord Shelburne—sat in opposition to these voluntaristic theologies. His critics argued that his monistic approach to spirits was "nothing better than a heretical ghost story," though it presumed to improve upon the superstitious religious beliefs of traditional Christianity. But even Priestley ultimately considered matter subordinate and inferior to divine power, and remained far more optimistic about the reality of causes than Hume. Like many of his colleagues, he also saw natural philosophy and true Christianity as allies in the battle against superstition and political oppression. He also shared with many of his less radical contemporaries the belief that expert philosophers and medical managers should run the state.[26] For all his hereticism, Priestley joined the British project of governing sectarian idiosyncrasy with rational but not arreligious policy.

Amidst all their opposition on the finer points, then, the various religious factions within the early industrial natural philosophical community maintained a consensus that one should not ascribe phenomena to divine causes too quickly. A strong center held in British natural philosophy, one that included a wide swath of scientific reformers on many other issues. On the one hand, French, Enlightenment-era science had explained many phenomena that had no explanations in Newton's day. The need for an intervening God had thereby diminished. On the other hand, moderate and conservative Britain recoiled from the Revolution and Reign of Terror, and the resurgent popularity of natural theology owed something to this revulsion.[27] Those who reveled in the empirical successes of recent French science, also cringed at statements such as this one from Julien la Mettrie's *Man a Machine:* "Admettre une *âme,* . . . c'est être réduit à l'*opération du St. Esprit*" (To admit the existence of a *soul* . . . is to be reduced to the *operation of the Holy Spirit* [translation mine]).[28] Thus, at the same time as British natural philosophers sought to reassert how the study of nature revealed its divine origins, they also hoped to accredit as little as possible in nature to divine interference. God was the first cause, but rarely the immediate one any longer.

Mainstream British natural philosophy both maintained a steady dualism (with respect to mind/body and God/nature) and used this to forge a neater, friendlier compromise between the realms of natural philosophy and theology. Fire and brimstone, excessive ritual, incantations—all seen as immoderate approaches to faith—violated both true religion and scientific rationality. Both science and religion thus benefited from meeting on moderate political grounds; in William Carpenter's words: "Science [gained] by being led to regard all the phenomena of Nature as manifestations of the constant and all-pervading energy of a Mind of infinite perfection;—Religion, by obtaining that expansion and definition of its ideas as to the unlimited range and predetermined order of the Divine operations, which the Scientific conception of them alone can afford."[29] To trade in subjective experience for general knowledge, both natural philosophy and religion had to abandon their more extreme positions.

SCIENCES OF THE UNSEEN

The delicacy of dealing with faith and causation in nature became manifest in what we might call the sciences of the unseen, those areas of natural philosophy that considered phenomena whose causes were not directly visible to the eye. I mean subjects as diverse as gravitation, matter theory, the sciences of the imponderables (electricity, magnetism, light, heat, and nervous impulses), mesmerism, and, later, spiritualism. In this brief section I do not intend to rehearse the already well-documented histories of these sciences. Since we are interested here in the ideas of natural philosophers on spirits, we must nonetheless take account of the major issues that arose in these various sciences. Early industrial natural philosophy had a complex relationship with other immaterial processes. The actual forces of magnetism and gravitation, for instance, were perhaps no more material than ghosts or the magnetic fluids running through the body of a mesmerized patient. Ever since accepting Newton's action-at-distance scheme of gravitation, natural philosophy had abandoned a strictly materialist vision. Rigorous rules of hypothetico-deductivism and experimentation effectively replaced materiality as a guarantee of the reality of immaterial phenomena. Natural philosophy, for instance, was supposed to concern itself only with "causes recognized as having a real existence in nature, and not being mere hypotheses or figments of the mind."[30] Deepening the fine line between "real existence" and nonsense mattered greatly to natural philosophers who needed to clarify the boundaries between the real and imaginary in their own minds and between the ordered and disordered in their changing culture.

The epistemology of causes in natural philosophy was therefore extremely contentious. Heated arguments grew in part out of theological differences, each disputant claiming to have found the best blend of reason, empiricism, and faith. The early-nineteenth-century controversies over the nature of light and the existence of the ether, for instance, illustrate deep divisions concerning hypothetical entities. Theological commitments encouraged men such as Brewster and Henry Brougham, for example, to dismiss the wave theory. While their colleagues saw the causes of natural phenomena as *metaphors* for God's operation on earth, Brougham took God's intervention more literally. To argue that light was a mathematical construct heretically implied that God was as well.[31]

The ether also served as a device for working out the relationship between God and the physical universe, mind and matter. Whether or not a particular natural philosopher considered the ether valid depended largely on his or her theology. Broad-church dualists tended to reject the possibility of active ethereal matter. "Mechanical" ethers appealed more to atheists and other materialists; while "animate" ethers found support among spiritualists.[32] There were other, interesting variations: believing in the manifest nature of creation, the evangelical Calvinist Brewster feared that allowing theoretical entities in natural philosophy might leave natural theology vulnerable. A natural philosophy based on facts would never become obsolete, and would thus remain impervious to religious skeptics.[33] Despite these dramatic differences, most British natural philosophers sought a matter theory that admitted some mechanism for divine intervention, a mechanism that left as much as possible to reason and empiricism rather than revelation.

Likewise, even by the mid-nineteenth century, natural philosophers were not unequivocal in their dismissal of mesmerism, just as they would later have mixed reactions to spiritualism. James Gordon and Augustus de Morgan were fascinated with mesmerism, to their friend John Herschel's chagrin.[34] Michael Faraday conducted his own investigation of spiritualists' claims to channeling electric, magnetic, diabolical, and supernatural agencies. He found them all wanting, and particularly deemed the last two causes to be "too much connected with credulity or superstition to require any attention on [my] part." His own evangelical Sandemanianism underlay his phenomenalist approach to nature, which led him to reject speculative claims about causes whether they applied to table-rapping or forces.[35] Maxwell attended a seance in Edinburgh during his Christmas vacation from Cambridge in 1850, but was disappointed by the operator's failure to make Maxwell forget his own name.[36] He later presented an essay at Cambridge in which he derided the "Dark Sciences" for their "speciously sounding laws of which our first impression is that they are truisms, and the second that they are absurd, and a bewildering mass of experimental proof, of which the tendencies lie on the surface and all the data turn out when examined to be heaped together as confusedly as the stores of button-makers."[37]

In every area of early industrial natural philosophy, the reality of theoretical, immaterial objects and occult causes aroused serious debate. For all of its epistemological differences and strictures, the metaphysics of natural philosophy nearly always left provisional room for the operation of strange, and occasionally even supernatural forces. Natural philosophers simply demanded that a phenomenon be given a natural explanation if one were available.

HALLUCINATIONS AMONG THE SANE AND SECULAR

In the period under discussion, hallucinations were divided into three classes: religious visions (left to theologians), insane imaginings (left to physicians), and "visions of the sane" (natural philosophers' domain). Hallucinations among the sane became a distinct possibility when public entertainments such as the magic lantern became popular in the late eighteenth century. The cause of these "normal" apparitions were understood to be external. In fact, many magic lantern entertainers proclaimed that one of their goals was to show the public the artificial nature of their own images. To Ezekiel Walker, inventor of the phantasmascope, optics taught nothing if not to distrust one's senses, "for when we use optic glasses, instead of viewing real objects, we generally see nothing but phantoms." Walker therefore saw it as his duty during his demonstrations to explain fully the mundane workings of his invention.[38]

While many in the early modern period believed that spirits had as external an existence as magic lantern projections, industrial-age natural philosophy construed hallucinations as internal productions of the mind. The illusions produced by magic lantern shows had some external basis, whereas true hallucinations did not. Physical factors such as illness might exacerbate them, but ultimately the mind replaced the supernatural as the cause of apparitions.

A key part of this change was the successful establishment in the early industrial period of the power of the mind to alter perception. Victorian natural philosophers may not have trusted the imagination, but they certainly felt the need to acknowledge and discipline it. Like the magic lantern, the mind could play tricks, but could also be fathomed as an ultimately predictable instrument. The banishment of apparitions to the mind also meant, though, that all thought came to seem ghost-like. As Terry Castle has put it, "There was now a potential danger in the act of reflection—a danger in paying too much attention to mental images or in 'thinking too hard.' One's inmost thoughts might at any moment assume the strangely externalized shape of phantoms."[39] Indeed, as early as 1841, one prominent author on the subject of hallucination suggested that only a difference of degree distinguished ideas from phantoms.[40] If one could successfully control hallucination and other products of the imagination, one could then defeat not only insanity, but also its philosophical counterpart, idealism.

The Imagination in Natural and Moral Philosophy

We can begin to sense the profound change that Romanticism and *Naturphilosophie* wrought on science when we consider the difference between empiricist views of the imagination in the seventeenth versus the nineteenth century. For Locke, madness was any false association of ideas not quickly set right by the reason. Ideas were either real or "fantastical"; they either corresponded to the real world or they did not. "It matters not what Men's Fancies are," Locke wrote, "'tis the Knowledge of things that is only to be prized: 'tis this alone gives a value to our Reasonings, and preference to one Man's Knowledge over another's, that it is of Things as they really are, and not of Dreams and Fancies."[41]

For the mid-nineteenth-century rational empiricist, this position was no longer entirely tenable. By then, too many famously intelligent people—people neither insane nor privy to revelation—had been deluded by false visions (about which more later). According to his own methodology, the empiricist had to face the facts: too many sane, educated people hallucinated in quite secular ways. Locke's rigid distinction between the reasoned mind and the mad one had to be wrong. Benjamin Brodie noted, for instance, that bodily illness without any attending mental disorder could also instigate hallucination. And though the patient knew rationally that the apparitions did not exist in the real world, yet they did not thereby disappear.[42] The early-nineteenth-century literature on hallucination thus constituted a rejection of the earlier Lockean and Addisonian idea that vision and the imagination were allied mediators between God and humanity. In post-Romantic British natural philosophy, the imagination became the fall guy, an untrustworthy faculty unfit for revealing scientific knowledge.[43]

We find a clear example of this transition in the development of the Common Sense school in Scotland. Among the earliest of that school, Thomas Reid sought to counteract the skepticism of those such as Locke and Hume, who argued (as Reid saw it) that direct knowledge of the world was impossible. According to Reid, Locke's ideas, the fundamental components of perception, were mere representations of the true objects. But for Reid, sensation did not produce representations, but *direct* knowledge of the world. Locke's ideas were mere "phantasms." Among his many critiques of his mentor, Thomas Brown feared that Reid had not dismissed idealism as neatly as he had supposed. Nothing in Reid's arguments prevented the external world from being a fiction of the mind's creation. Taking dreams and madness to be an *experimentum crucis* for Reid's "proof" of an external world, Brown posed the rhetorical question: "Is it utterly absurd and ridiculous to maintain that all the objects of our thoughts may be 'such stuff as dreams are made of?' or that the uniformity of Nature gives us some reason to presume, that the perceptions of maniacs and of rational men are manufactured, like their organs, out of the same materials?" Something more than mere "common-sense" faith in realism would be required to distinguish the phantasms of maniacs from everyday perceptions of the world.[44]

The most popular answer in early-nineteenth-century natural philosophy was that training the intellective faculties would allow them to tame the imagination, the key culprit in hallucination. The imagination often interfered with true sensations of external objects, and could even make the mind think it perceived objects that did not actually exist. Dugald Stewart, for example, noted the power of the imagination to cloud perception and memory, and deemed the "vulgar" particularly inept at judging the difference between reality and fantasy.[45] Overestimating the imagination's importance might even lead to the Humean, phenomenological conclusion that no necessary connection existed between our perceptions and a putative "real world." Most British natural philosophers found this patently ridiculous. To John Herschel, for instance, denying cause-and-effect relations would reduce the "world of realities" to "a perpetual kind of sensible phantasmagoria from the recollection of whose images the mind might amuse itself in forming pictures and patterns but only as toys in which reason had no part." He also compared the idealist notion that the "mind can create for itself a universe" to "*Clairvoyance* which can see all that it knows *can* be seen but which never reveals anything new — never originates a discovery."[46]

In sum, the imagination could prove useful and creative in the appropriate contexts, but faculty psychology from the turn of the nineteenth century proclaimed the need to subordinate the imagination to the other faculties. Otherwise, one lost the basis for distinguishing facts from hallucinations, and good science from pure fancy, manageable abnormality from sheer madness, and enthusiasm from a religiously ordered nation.

Early-Nineteenth-Century Hallucination Literature

Building on these insights about the imagination and reason, a popular scientific literature on hallucination proliferated during the industrial age. Since the authors of these texts sought not only an audience of physicians and metaphysicians, but laypeople as well, their work had a broader reach than the material considered in the previous two chapters. These works expressed a broad-based consensus in early industrial Britain that explaining hallucination required a commitment to both reasoned empiricism and divine power.

The word *hallucination* first entered the English language through a 1572 translation of Ludwig Lavater's *Of Ghostes and Spirites Walking by Nyght*. Lavater considered hallucination to be any "strange noyses, crackes, and sundry forwarnynges, whiche commonly happen before the death of menne, great slaughters and alterations of Kyngdomes."[47] The physician Thomas Browne famously used the concept in his 1646 *Pseudodoxica Epidemica* to refer to what one's vision produced if it were "depraved and receive[d] its objects erroneously." The eighteenth century, as previously mentioned, increasingly medicalized hallucinations. Daniel Defoe's 1727 essay on apparitions is often considered a transitional work between the traditional, supernatural interpretation of apparitions, and the newer approach to these phenomena as the products of diseased bodies and vulnerable minds.[48]

Supernatural explanations for hallucinations did not disappear entirely, however. In 1803, Elias Carpenter, minister to the millenarian "House of God" near Newington Buffs, published his *Nocturnal Alarm*. At the time, Carpenter was a follower of the notorious medium Joanna Southcott, the so-called "prophetess of Exeter," though he later renounced her. As the responses to Carpenter and Southcott made manifest, their position had long ceased to be a viable one for most natural philosophers.[49] Still, as late as 1856, an originally skeptical wine merchant named Newton Crosland became convinced of the supernatural communications he experienced at table-rappings. He declared that it was both more scientific and more devout to grant the overwhelming testimony of spiritualist witnesses. A more intimate acquaintance with the Victorian scientific community would have shown Crosland that he had more sympathizers with his basic precepts than he knew. Many of those who did not believe in table-rapping still would have acknowledged that dogmatic materialism violated the very ethos of natural philosophy. Many would have joined in his cheers for "Protestantism . . . enlightener and civiliser of the world! yet Protestantism, ignorantly and materialistically, shrinks from owning the miracles wrought within her own pale. She vacantly stares at the pillar of flame offered for her guidance, and sceptically calls it an *ignis fatuus;* she childishly resigns one-half of her inheritance of light, and hands it over to less worthy keeping."[50] Despite their general tenor of skepticism, early industrial natural philosophers for the most part continued to view rational, nonsectarian, Protestant faith as the key to general knowledge and invisible forces as plausibly real entities.

The more accepted literature on hallucination at the turn of the nineteenth century turned its attention to demarcating sensory from mental errors. Many physicians continued to run together the categories of hallucination and illusion, and considered them both reliable tools for diagnosing insanity. For example, Alexander Crichton used hallucination and illusion interchangeably, though he feared that a considerable number of sane people were being misdiagnosed for perfectly innocent optical and mental mistakes.[51] Thomas Arnold differentiated totally fabricating an object from grossly misperceiving an object that actually existed, but called them both forms of madness.[52] Such work tightly wove the study of hallucination into the larger fabric of treating insanity. But in 1838, Etienne Esquirol clarified the highly important distinction between illusions and hallucinations. The former had some external referent, however misjudged; the latter had none. Long thought to have some neurological basis, the "pretended sensations of the hallucinated are images and ideas reproduced by memory, improved by the imagination, and personified by habit."[53] While Esquirol's work did not preclude a materialist reading of hallucinations, British writers tended to attribute that type of vision to the workings of the memory and imagination rather than the physical brain.

Before Esquirol's breakthrough, a number of physicians and natural philosophers had begun to look for ways to understand hallucination that did not necessarily entail madness. One of the most important new works in this respect was John Ferriar's 1813 *Essay towards a Theory of Apparitions*. As senior physician to the Manchester Infirmary, Lunatic Hospital and Asylum, Ferriar sought to encourage the new trend toward treating apparitions as mental products. He marshaled numerous case studies — historical and current — to show that "certain states of the brain" were the main precondition for these visions. Those mental states could be brought on by, for instance, poor health or melancholy. By way of example, he pointed to the "great prevalence of spectral delusions" during the Interregnum, when the "melancholic tendency of the rigid puritans" dominated the national mood. These delusions threatened the very integrity of the political fabric. Ferriar stressed that hallucinations did not occur only to the insane; he merely sought to strip them of their supernatural associations. "Instead of regarding these stories with the horror of the vulgar, or the disdain of the sceptic, we should examine them accurately, and should ascertain their exact relation to the state of the brain, and of the external senses." Toward this end, experimental philosophers were to take just as much interest in the problem as physicians.[54] Not only bodily therapeutics, but also natural philosophy's understanding of the mind could help contain a problem that had the potential to cause as much social disorder in the industrial age as it had during the English Civil Wars.

Like many others who wrote about hallucination, Ferriar found Christoph Friedrich Nicolai's experiences helpful for explaining how the methodical investigation of visions might work. A bookseller in Berlin, Nicolai had published several famous commentaries on Goethe and Schelling. Hardly a believer in ghosts, however, he dismissed the objective reality of apparitions and the good sense of the people who believed in them. In order to compound the evidence against the idea of phantasms as real things, Nicolai offered his own visions to the Royal Society of Berlin. In 1791, after a physically and emotionally difficult period, he "saw, in a state of mind completely sound, and after the first terror was over, with perfect calmness, for nearly two months, almost constantly and involuntarily a vast number of human and other forms, and even heard their voices, though all this was merely the consequence of a diseased state of the nerves and an irregular circulation of the blood." Nicolai claimed that both he and his physician immediately recognized his visions as illusions brought on by his disturbed nerves, though they were confused when the apparitions continued despite his improving condition. In fact, their frequency increased.[55]

Though he could not control the appearance of the visions, Nicolai claimed he could remain calm enough to observe them carefully and recognize their difference from real objects.

> Had I not been able to distinguish phantasms from phenomena I must have been insane. Had I been fanatic or superstitious, I should have been terrified at my own phantasms, and probably might have been seized with some alarming disorder. Had I been attached to the marvellous, I should have sought to magnify my own importance, by asserting that I had seen spirits; and who could have disputed the facts with me? The year 1791 would perhaps have been the time to have given importance to these apparitions. In this case, however, the advantage of sound and deliberate observation may be seen. Both prevented me from becoming either a lunatic or an enthusiast; for with nerves so strongly excited, and blood so quick in circulation, either misfortune might have easily befallen me. But I considered the phantasms that hovered around me as what they really were, namely, the effects of disease; and made them subservient to my observations, because I consider observation and reflection as the basis of all rational philosophy.[56]

Nicolai's commitment to the generalization of subjective experience is palpable. Only by finding organic and rational causes for his visions could he hope to enter the scientific discussion he sought. In fact, simply stipulating his commitment to empiricism constituted the most important therapeutic step in fighting back madness and religious enthusiasm. The application of leeches simply finished the cure. It is easy to see why this case appealed to Ferriar: In the midst of his mentally disturbed state, Nicolai claimed he never once suspected his visions had a supernatural origin, but rather searched for cause and cure in his own physiology and morals. The bookseller's case supported the point that hallucination had a scientific explanation, and could occur to those whose nervous system was not irrevocably disturbed.

One of the many besides Ferriar who found Nicolai's tale compelling was naturalist and antiquarian Samuel Hibbert. An active participant in the scientific communities of Manchester and Edinburgh, Hibbert had access to many such cases, both published and unpublished. He concluded, like Ferriar and Nicolai, that hallucinations had natural origins and often occurred to sane people.[57] Hibbert carefully left open the possibility that God may have revealed himself in the post-apostolic age, but deemed it highly unlikely. Instead, inspired by the success of Nicolai's leeching and other such cases, he looked to changes in the circulatory and nervous systems.[58] Knowledge of this physiology was still imperfect, but the potential for certainty and universal consensus far exceeded that in theology or the psychological understanding of the imagination. Nicolai's reasoned approach to his illness and visions, Hibbert believed, proved that imagination despotically ruled neither the production of the images nor the study of that production.[59] Hibbert's strategy for controlling the imagination was thus to deny its power.

In his own writing on hallucination, John Abercrombie also remained skeptical of the imagination's power. The son of an Aberdeen minister, Abercrombie had replaced Samuel Gregory as the leading consulting physician to Edinburgh and, effectively, all of Scotland. Though his books on pathology and his introduction of vaccination to Scotland improved his reputation, his *Inquiries concerning the Intellectual Powers* became his most famous work. He found too speculative the Common Sense philosophers' and phrenologists' ideas about mental faculties. Still, he recommended that reason and judgment (whatever their metaphysical status) be cultivated as guides toward true knowledge. They would help one walk the middle path between credulity and skepticism.[60]

Through his own medical practice in Surrey, William Newnham similarly found material for his popular medical and religious ideas. A devoutly religious man who liberally peppered his writing with his faith, Newnham first published his ideas on hallucination in the *Christian Observer*. He nevertheless thought that most, if not all, modern visions had physical rather than supernatural origins. Organic diseases in the brain led to disorders in the mind. Likewise, moral profligacy could cause brain dysfunction. In the end, all apparitions had some basis in material processes.[61] This view, he argued, did not make him a heretical materialist. On the contrary, the only path to Christian revelation lay through the constant exercise of reason. Belief in supernatural visitations violated both reason and Scripture.[62] One must never forget God's place as the First Cause, he admonished, but realistically one could only hope thoroughly to grasp secondary causes, all accessible via reason and observation.[63] He hoped to bring together scientifically minded Christians of all stripes (with the possible exception of the supposedly overzealous and irrational Catholics) to battle against enthusiasts and skeptics.[64] Once men of science experienced faith in its more rational form, "he is assured that religion is not that contracting study which he once thought it, but that it possesses the power even of ennobling the mind, and thus the veil of prejudice is blown aside, the film of visual delusion dissipated, and at least the soil is prepared for the reception of Divine truth." Here, a rational, nonprovincial faith explicitly played the role of curing hallucination.[65] It also warded off the kind of anarchy brought by excessive skepticism or enthusiasm. In a metaphorically pregnant passage, Newnham defined hallucination as the state of the brain when it was "no longer the obedient servant of the mind; but, in the tyranny of usurpation, subjugates the reasoning powers, and compels them to yield to that human infirmity, which attaches itself to the grand prevailing cause that has marred the most perfect creation of Omnipotence, and has rendered that which was originally 'very good,' now 'very far gone from original righteousness.'"[66] "True religion," or that guided by reason, offered just the mixture of moral and physical remedies needed to stem the tide of mob irrationality that threatened order in Britain just after the French Revolution.

Walter Cooper Dendy, senior surgeon to the Royal Infirmary for Children on Waterloo Road in London, also weighed in against the dual terrors of superstition and radical materialism. Too many people, he said, made the mistake of thinking that to deny one position was to affirm the other. Dendy preferred a middle road on which the natural world could be studied without being stripped of either its spiritual meaning or its reliable physicality. In his *Philosophy of Mystery*, a dialog among four young people, Evelyn (Dendy's avatar in the conversation) professed his unwillingness to "blend the sacred truths of spiritual futurity with arguments on the imperfection of material existence." A deistic philosophy, Evelyn argued, more closely approached the Deity than either skepticism or superstition. Indeed, the only evidence of supernatural influence in the modern world was the mind, "which in its [original,] pure state was itself an emanation from the deity."[67] Anatomical research, combined with a rigorous, unsentimental metaphysics, would yield the best possible knowledge of the ultimately ineffable mind. While the other three characters felt varying levels of nostalgia for the days of fairies and demons, Evelyn/Dendy expressed a common, loosely deistic sentiment that a world created and left to run alone by God was wondrous enough.[68] The belief in demonic possession and other supernatural effects, Dendy's conversationalists agreed, persisted mainly among "backward" people such as the Irish, the Highland Scots, Eastern Europeans, and various "savages" elsewhere in the world.[69] Seeing phantoms was common enough—Dendy himself seems to have had a brush with them[70]—but the modern faithful dismissed such experiences as unreal.

From this substantial literature on hallucination, we must single out two further works: those by Sir Walter Scott and David Brewster, the first being the direct inspiration for the second. Their work crystallized the point that natural philosophy found hallucination an important subject because of its implications for the rational control of religious factionalism and other idiosyncratic phenomena.

Scott and Brewster on Natural Magic

Both Scott's and Brewster's volumes appeared as part of the inexpensive Family Library series published by John Murray and backed by the Society for the Diffusion of Useful Knowledge (SDUK). Murray and SDUK designed the series to spread reasoned thinking skills among the working classes. J. G. Lockhart (editor of the *Quarterly Review*) asked his father-in-law, Sir Walter Scott, to write on superstitions.[71] The by-then famous author honored the request, though it meant writing while nearly blind and just after a severe "paralytic seizure" in February 1830.[72]

Characteristic of his Romantic but critical view of Scotland's past, in his *Letters on Demonology and Witchcraft*, Scott both scorned and marveled at the credulity of medieval and early modern Europeans. Rampant belief in witchcraft and astrology, fairies and ghosts, he thought, revealed the undisciplined imagination endemic to immature civilizations. Even the most intelligent people could not fully escape the gullibility of their times. Then, the founding of the Royal Society and other scientific institutions in the mid-seventeenth century set society on the course of reason and common sense. Over time, more and more phenomena succumbed to natural law. Each discovery thus supported the idea that God governed the universe through fixed laws and only rarely through miraculous intervention. By his own time, people far less frequently fell prey to obvious superstitions, such that Scott mused that his own rather basic treatise might already be out of date: "Even the present fashion of the world seems to be ill suited for studies of this fantastic nature; and the most ordinary mechanic has learning sufficient to laugh at the figments which in former times were believed by persons far advanced in the deepest knowledge of the age."[73]

Like an infant society, the young in any era were more susceptible to superstition than their more circumspect elders, a truth Scott knew firsthand.

As a young man, he had spent a night in the Earl of Strathmore's castle, a romantic locale that easily inspired imaginative reveries. Alone in his room that night, he could not banish thoughts of the murderous scenes in *Macbeth*. "In a word," he wrote, "I experienced sensations which, though not remarkable either for timidity or superstition, did not fail to affect me to the point of being disagreeable, while they were mingled at the same time with a strange and indescribable kind of pleasure, the recollection of which affords me gratification at this moment." He had even dabbled in demonology in his twenties and thirties.[74] In his more sensible (or at least wearier) forties, Scott again spent the night in a haunted castle (Dunvegan, in the Highlands), but this time found the comfortable bed a more "engaging spectacle" than any supposed ghosts or goblins.[75]

By discrediting the supernatural, Scott did not wish to deprive people of their sense of wonder, for "the abstract possibility of apparitions must be admitted by every one who believes in a Deity, and His superintending omnipotence." He considered it quite normal for educated people to experience the occasional hallucination. Rather, he sought to congratulate his contemporaries on not confusing the productions of their fancy with the rare miraculous fact. Widespread education and common sense gave industrial society the power to determine the illusory quality of virtually every one of these sightings. One could examine oneself and one's circumstances to find the source of these illusions: for instance, illness, overactive imagination, or deliberate deception by another party.[76]

While they did not have exclusive claim to reasoned judgment, natural philosophers modeled good behavior in this regard. For example, while attending a meeting of the Berlin Academy, Scott heard German botanist J. G. Gleditsch tell his story about seeing the figure of Pierre-Louis de Maupertuis, the recently deceased French geographer who had long served as president of the Academy. Gleditsch, convinced the sighting had been a mere trick of his eyes, told his colleagues the story as a mere curiosity.[77] Scott related a similar story about the members of a science and literature group in Plymouth. At one of their meetings they thought they saw an apparition of their critically ill president. Upon sending two members to the president's house to seek a reasonable explanation, they learned from his nurse that the sick man had died that very evening. Too "philosophical" and "wise" to think they had seen a ghost, the society's members kept the story secret until they learned several years later that the nurse had lied. The president had walked over to the meeting in a stupor before being retrieved by the nurse. He had died afterward. Scott noted that "the philosophical witnesses of this strange scene were now as anxious to spread the story as they had formerly been to conceal it, since it showed in what a remarkable manner men's eyes might turn traitors to them, and impress them with ideas far different from the truth."[78]

Scott also told the strange tale of a gentleman of legal and scientific attainments, who during a severe illness came under the care of a physician friend of Scott's. The scientific gentleman was so plagued by hallucinations that he feared his reason was "totally inadequate to combat the effects of my morbid imagination, and I am sensible I am dying, a wasted victim to an imaginary disease." He occasionally saw a large cat, which he first thought to be one of the household pets, but eventually realized was the product of his "deranged visual organs or depraved imagination." The cat was replaced by an apparition of an usher, who constantly attended him. This vision in turn gave way to one of a skeleton. He knew the apparition could not be real, and yet it remained. The physician tried to employ the man's superior powers of reason and common sense as a self-cure, but his imagination had grown too strong. Scott marveled at the power of the imagination to "kill the body, even when its fantastic terrors cannot overcome the intellect."[79]

Encased in these stories, and in *Letters on Demonology and Witchcraft* as a whole, was the warning that the intellective powers had to maintain a constant vigil over the imagination. Fancy could serve useful moral purposes, but undisciplined, the imagination quite easily led the most sensible man into folly, even death. While Scott believed that natural philosophy had helped make the nineteenth century a less superstitious age, each rejection of a hallucination as illusory was still an achievement, a small victory for human self-control over the idiosyncratic excesses of superstition.[80] Constant vigilance necessitated permanent leadership by cooler heads.

When Scott published his *Letters*, he was in frequent contact with his fellow countryman David Brewster. Soon after, Brewster obliged Scott's request that he write a "popular account of those prodigies of the material world which have received the appellation of *Natural Magic*."[81] Brewster intended his book to supplement rather than supersede Scott's, and he accomplished this by concentrating on his own area of expertise, optics. More than any other science, he wrote, optics was "the most fertile in marvellous expedients," something Brewster well knew as an inventor of the kaleidoscope, one of the most wildly popular optical toys of the industrial age. Vision had the most expansive powers of all the senses, he wrote, but for that reason also had the greatest capacity for error. The mind imputed so much importance to vision that it often, literally, projected its own ideas on the retina: "The optic nerve is the channel by which the mind peruses the handwriting of Nature on the retina, and through which it transfers to that material tablet its decisions and its creations. The eye is consequently the principal seat of the supernatural."[82]

Brewster demonstrated the mundane nature of many "supernatural" visions by telling stories, as Scott had done, and by inviting his readers to perform simple experiments. To grasp the truth behind these demonstrations, the reader had to come to them without fear or prejudice. She had to be willing to accept controlled ocular demonstrations as proof of the reality-altering power of the imagination. The imaginative construction of apparitions was also aided by various kinds of optical illusions, where the eye was tricked without any assistance from the mind. Illness, an overactive imagination, emotional excitement, or distraction only compounded the effect. These illusions—all easily explained—could fool even an intelligent observer in perfect health, as in the case of Wollaston's hemiopsy. "When a phenomenon so strange is seen by a person in perfect health, as it generally is, and who has never had occasion to distrust the testimony of his senses, he can scarcely refer it to any other cause than a supernatural one." This impulse to believe apparitions had supernatural causes was psychologically understandable, but the better solution was to find a rational explanation, as Wollaston had.[83]

Brewster also approvingly reported the case of a Mrs. A., a model for his readers on how to handle the onset of hallucination. In a poor state of health, Mrs. A. alternately saw the ghosts of her dead sister-in-law, her husband, and a cat. What made her case so remarkable was her extreme perspicacity and educated good sense, both authenticated by her social station. These qualities in turn ensured the trustworthiness of her account and enabled her to investigate the true nature of her visions. Brewster's admiring evaluation recalls Herschel's view of Somerville discussed in chapter 2:

> The high character and intelligence of the lady, and the station of her husband in society, and as a man of learning and science, would authenticate the most marvellous narrative, and satisfy the most scrupulous mind, that the case has been philosophically as well as faithfully described. . . . From the very commencement of the spectral illusions seen by Mrs. A. both she and her husband were well aware of their nature and origin, and both of them paid the most minute attention to the circumstances which accompanied them, not only with the view of throwing light upon so curious a subject, but for the purpose of ascertaining their connection with the state of health under which they appeared.[84]

Rather than believe her visions supernatural, Mrs. A. recalled a "crucial experiment" for telling ghosts from illusions that Brewster had once suggested to her. She pressed her eye until she saw double, after which everything in the room but the "ghost" appeared double.

By admiring her intelligence and social standing, Brewster did not mean to limit intelligence to the wealthy classes. In fact, he hoped to extend their advantages to the "great mass of society," who ignorantly fell under the spell of mesmerists, spirit rappers, and diviners. Mrs. A.'s position guaranteed her veracity, but it did not grant her exclusive rights to it.[85]

The cases of Mrs. A., Wollaston, and other persons "of sound mind" who saw "spectral apparitions in the broad light of day" presented a key irony for Brewster. It seemed inexplicable that reasonable people could fall prey to such bold tricks of the mind. He speculated that such "persons of studious habits" might become so engrossed in abstract thought that "external objects even cease to make any impression on the retina." This theory gained credence through circumstances such as Mrs. A.'s, who was deep in concentration on a "striking passage in the Edinburgh Review" when she first saw her sister-in-law's ghost. Even Samuel Hibbert, on a visit to Brewster's home, was startled by what he thought was a ghost in his chamber (which turned out to be the glow of his nightcap, which had caught fire).

Brewster himself remarked to his friend Mary Somerville that though he did not believe in ghosts, he was "*eerie* when sitting up to a late hour in a lone house that was haunted."[86] His daughter Margaret Maria Gordon claimed that Brewster's upbringing in Jedburgh left him "suffering from superstitious fears even up to the mature years of manhood."[87] He found useful the strange fact of intelligent people hallucinating; through these otherwise careful and reliable witnesses, "yet we may arrive at such a degree of knowledge on the subject as to satisfy rational curiosity, and to strip the phenomena of every attribute of the marvellous."[88] Margaret Gordon's comments about her father's fears and Brewster's own anti-elitism suggest that he expected not just the masses but also men like him to take comfort in others' near-escape from folly. Brewster never meant his *Letters* as an admonishment to the witless.

Brewster hoped to engage all of his readers, educated or not, in a project of stripping these phenomena of their supernatural qualities. He spent most of the book explaining various kinds of visions seen in a normal state of health, since most intelligent people could independently detect the deception behind the hallucinations of illness and carnival tricks. For instance, the Brocken specter seen in the Harz Mountains in Hanover was a favorite vision of Romantic writers. At a certain place on the summit of the Brocken would appear a ghostly form, several hundred feet in height, which would mimic every move made by the observer. As Haue had first realized while visiting the place in 1797, this figure was simply a reflection of the observer created by the peculiar atmospheric effects of that spot. The mountainous region, once so full of wonder and horror, became for midcentury rationalists little more than "the grand temple of Saxon idolatry."[89]

Brewster admitted that such explanations were necessary to eradicate our quite forgivable sense that a supernatural agency lay behind hallucinations. As a Calvinist, he opposed attempts to lash science to Scripture. Religion, for all its moral authority, should not hinder free scientific inquiry. Pursuing rational explanations made one more, not less, knowledgeable of God's creation. In other words, to deny supernatural agency in hallucinations was not to deny divine power. Quite the contrary: scientific explanation affirmed God's power by denying demons. As with his opposition to the ether, Brewster rejected metaphysical explanations for hallucinations as unscientific and too ephemeral to link to the bedrock of Christian faith.[90]

SANE HALLUCINATION AS A PSYCHOLOGICAL ISSUE

Scott and Brewster had roundly rejected divine communication as the source of modern hallucination. Yet they also dismissed purely materialistic explanations. Hallucinations were not simply illusions that could be wholly accounted for by physiological causes. Judgment, character, memory, the will, and imagination also contributed to the production and understanding of hallucinations. In effect, the mind replaced God and the supernatural as the intervening, immaterial cause behind these visions. To make sense of these experiences therefore required both physiology *and* mental-moral philosophy. For instance, one physician noted of a patient of his that he was "a man extensively acquainted with physiology; but felt utterly at a loss to what derangement, of what part of the animal economy to refer it. So, indeed, was I."[91] It became increasingly clear during the industrial age that some new hybrid of physiology and mental-moral philosophy was needed to understand human mental life. This hybrid eventually became the professional discipline psychology.

The psychologizing of hallucination grew out of the literature of the first three decades of the century, but achieved full instantiation in Alexandre Brierre de Boismont's work in the 1840s. Brierre de Boismont, who had been treating the insane in Paris since 1825, recognized in his own work and the previous literature a need to claim hallucination for psychology. Hallucination, he argued, sprung from causes, and responded to treatments, that materialist medicine alone could not address. He drew on Esquirol's distinction between illusions and hallucinations to assert that neurological systems such as David Hartley's could explain sensory illusions, but some visions had no apparent basis in the nervous system. Against the prevailing materialism of his Parisian medical colleagues, he rejected "the opinion which can only see in lunacy a pure and simple nervousness, like chorea, hysteria, and epilepsy; and in reason the product of a physiological action entirely material. To us, ideas have a different nature than sensations. Psychological facts cannot be placed on the same line with those that affect the senses. Although the brain may be the seat of intellectual operations, it is not the creator of them. The notion of the idea exists before that of its representation."[92] Materialist pathology alone could not explain how the mind's imaginative faculty could continue to produce hallucinations even when the reason knew they were false. Hallucination could even inspire people to great deeds.[93]

Since Brierre de Boismont believed that neither physiological laws (by themselves) nor the supernatural explained hallucination, he thought it important to draw a three-way distinction between faith, intellect, and matter. This allowed him to reach several conclusions important to dualist psychology. First, he could qualify the kind of distinction that Esquirol had drawn between hallucination and illusion. Hallucination consisted of the seeming perception of objects originating in the mind. Illusion resulted from the misperception of actual objects. Hallucination and illusion had analytically distinct etiologies, but this masked the more important fact that mind and body *both* contributed almost seamlessly to phenomena such as hallucinations. By approaching hallucination as a psychological issue, one could gain a fuller picture of its causes, both moral (such as overindulgence in fairy stories as a child, lack of education) and physical (like nervous disease or the consumption of drugs or alcohol). This approach could also explain the occurrence of hallucination in the sane, without forcing one to resort to phenomenology.[94]

Second, he could explain that for a healthy psychology, one needed to balance faith and reason. "Faith without reason leads directly to superstition," argued Brierre de Boismont; "and reason without faith almost always results in arrogance. The hallucinations which arise from these two sources of error will be as various as the habitual ideas and occupations of the individual." Ideologues of any stripe—whether religious, spiritualist, materialist, or rationalist—each had their own distorted view of the world. Put another way, each hallucinated. For example, Brierre de Boismont believed it was the most rational, yet also the most faithful, position to believe that some Biblical-era people indeed saw or heard God. Since then, however, those claiming divine communication were almost certainly "madmen, dupes, and impostors." Brierre de Boismont breathed a sigh of relief that demonology had "greatly diminished since the 18th century."[95]

In this scheme, which played very well in reform-era Britain, the supernatural not only played a diminished role in nature, but the perfection of God's creation also came into question. The body—including the much-admired eye—was now known to have a number of flaws. Eighteenth-century natural theologians such as William Paley had admired the eye as one of the most exquisite of God's creations.[96] Concentrated study of the anatomy and physiology of vision in the early nineteenth century, however, revealed that the very structure of the visual apparatus could interfere with perfect vision. The fluid needed to lubricate the conjunctiva caused halos and false streams of light. The lashes required to protect the eye could be mistaken for objects in a microscope. Having two eyes helped one correctly judge three dimensions, but could also cause specters when their movements did not correspond. "It seems," reflected James Jago in 1854, "to result in Nature's works, as in man's, that no advantage can be consummated without entailing an appreciable disadvantage."[97]

What sustained faith in God's grace was the power of the human mind to correct for or transcend these flaws. Psychology thus gained in importance. In his Bridgewater Treatise, Thomas Chalmers thought that one could find evidence for design in the harmony between the mind's subjective concepts and objective nature. And in Herschel's view, God's creation remained incomplete without man's "intelligent worship" comprehending the universe's "INEFFABLE PURPOSE."[98] The psychological approach achieved by Brierre de Boismont and others helped to seal the conviction in the industrial age that what the imperfect body took away, the well-kept mind gave back. The literature on hallucination had established that spirits did not generally communicate directly with humans through apparitions, and the physiological and natural theological literature had admitted the imperfections of the human frame. What enabled all of these studies to continue was the optimism that the mind overcame all these shortcomings and disconnections from the world beyond. Inspiration for this optimism came from the fact that many great scholars had themselves hallucinated, and had made rational sense of their experience.

Natural Philosophers and Their Apparitions

In addition to their scattered allegiance to various sciences of the unseen, natural philosophers were also no strangers to hallucination. The experience of hemiopsy, as described in the previous chapter, was disconcerting enough. But when those visions began to take on the shapes of real objects, such as friends or landscapes, they became downright uncanny. As we saw in Brewster's case, natural philosophers' attempts to rationalize hallucinations were not directed only at the gullible man on the street or woman of the parlor. Many men of science in the early industrial period had their own experiences with apparitions, and again saw their efforts to heal themselves as part and parcel of a larger project to heal their nation.

Scholars—natural philosophers included—had long been thought susceptible to visions. Their supposedly characteristic melancholy, sensitivity, reflectiveness, isolation, and piety all contributed to this tendency. Several famous early modern cases reinforced this belief. In the mid-sixteenth century, Girolamo Cardano, a medical astrologer and mathematics teacher in Milan, frequently fell into trances. During these trances he felt his soul separate from his body, and he became virtually insensible to sounds or pain. He imagined he saw castles, houses, animals, and other objects.[99] Michael Mercato had seen an apparition of the late Platonic humanist and alchemist Marsilio Ficinus.[100] In 1610, half-asleep after his fatiguing attempts to understand the soul, Jean Baptiste van Helmont envisioned a table on which stood a bottle filled with liquid, but he was too horrified to drink it. A voice asked him if he wished for honors and riches. He then saw a great light through a slit in the wall. His passion to understand the soul continued, and in 1633 he had a second vision wherein he understood that the light in the wall had been his own soul.[101] Pierre Gassendi had advised his patron, the Count d'Alais, that a specter the count had seen was a visual or atmospheric effect, but insinuated that "perhaps it might be sent from Heaven to him, to give him a warning in due time of something that should happen."[102] Nicolas de Malebranche heard the voice of God within him. After a long seclusion, René Descartes found himself followed by an invisible person who pleaded with him to continue his philosophical research. Blaise Pascal, while working on the mathematics of the cycloidal curve, imagined that an enormous pit had opened in the floor next to his desk. Though he knew it was not real, Pascal still strapped himself to his chair while the illusion lasted.[103] A specter occasionally visited Lord Byron, though the latter chalked the experiences up to overexcitement of the brain. Samuel Johnson once heard his mother call him, though she lived far away.[104] Alexander Pope saw an arm emerge from the wall during a doctor's visit.[105] Johann Wolfgang von Goethe once saw a spectral doppelgänger approach him.[106] Pierre Cabanis also had once seen spectral "people walking on a pathway at a distance."[107] And Swiss naturalist Charles Bonnet, whose poor eyesight had forced him to give up use of his microscope, later in life began to see apparitions of men, women, birds, and buildings. They were the very same kind of experience he had observed years before in his grandfather, who had been equally sensible about recognizing the unreality of his visions.[108]

Similar experiences occurred to men of science in the early industrial period as well. Johannes Müller had noted that the eye could be stimulated either externally or internally by physical malfunction, memories, or the imagination. He could easily call up images to his own eyesight, and suspected that a more gullible mind might think these images to be divine or magical.[109] Hanwell Asylum manager John Conolly had treated an anatomy student who believed that an entire town lived inside his deltoid muscle. A physician friend of Walter Cooper Dendy's had treated a "man of great science" who received regular visits from phantoms. At first the man of science enjoyed his phantoms' company and so did not seek to stop the visions, but they eventually turned into "demoniac fiends, uttering expressions of the most degraded and unholy nature."[110] French physiologist Claude Sandras, who had written a treatise on nervous disorders, and German *naturphilosophische* physiologist Franz von Paula Gruthuisen reported experiencing hallucinations.[111]

A couple of reports connected seemingly supernatural phenomena to the influence of imponderable forces. In 1846, François Arago, Ernest Laugier, and Claude Goujon investigated the case of fourteen-year-old Angelique Cottin, whose body seemed to electrify everything it touched. They concluded that under certain conditions the human body could discharge a physical power and enact a kind of physiological action-at-a-distance.[112] In discussing the case some years later, an English writer lauded the three Academicians for their physical explanation, even though the phenomena seemed spiritual at first blush. Who, he asked rhetorically, "does not see, on sober reflection, that this hastiness [in attributing hallucinations to supernatural causes] does violence both to reason and to nature, and hence to their Founder?... There is as much sin in believing too much as in not believing enough."[113] Similarly, a German professor and hypochondriac, "M. J.," had sought help from Brierre de Boismont. His long hours working on magnetism, his abdominal pains, and the pressures of being educated beyond his originally low social station, had all conspired to make M. J. hallucinatory. Brierre de Boismont noted that his patient had the "fixed idea . . . that his friends injured him, placed him under magnetic influence, and that finally they had introduced a magnetizer into his abdomen." M. J. found relief from his visions through blisters, better nutrition, and the "judicious occupation of his mind in the analysis of important works."[114]

Like M. J., a young Swiss gentleman found relief from his visions through friends' aggressive attention to his diet and mental occupation. The young man one day fell into a stupor after having steeped himself in an intense study of metaphysics for six straight months. Though his actual senses seemed unaffected, his mind registered no sights or sounds, and he led his life as an automaton for a year. Eventually, a friend realized that reading to him very loudly seemed to restore some of his hearing. Each of the young metaphysician's senses were thus gradually restored (see Figure 5.2).[115] The Paris *Gazette Literaire* reported that a young mechanics student's zeal had led to sensory overload rather than deprivation: he bound himself tightly with ropes and hung himself from the ceiling. His pleasure from the experience ceased only when he realized nearly a day later that he could not extricate himself. In commenting on the incident, Hibbert noted the similarities of the man's erotic feelings to the ecstasy of those under religious persecution.[116]

In each of these cases, what might have been interpreted previously as religious ecstasy, British natural philosophers now interpreted unequivocally as mental and physical exhaustion. As we have already seen, various combinations of illness, fatigue, and lack of mental control could aggravate a person into nervous distress. After a raging fever in 1795, for example, an anonymous writer to Nicholson's *Philosophical Journal* became plagued by a number of hallucinations. Though he found his mathematical and other rational faculties undamaged, his dreams were haunted by enormous machines. He found some relief in being bled, training his mind to snap out of his feverish reveries, and drinking a cocktail of lemon juice, potash, and *pulvis Londinensis*. Still, one morning a parade of faces "over which I had no control" appeared to him and vanished.[117] Also while in the throes of a bad fever, London physiologist John Bostock saw figures that reminded him of those he had read about in Nicolai. Even after his delirium had passed, he continued to see the forms of a human face, a crowd of tiny figures, and other apparitions. Since he did not recognize any of the figures, he supposed his imagination responsible rather than his memory.[118]

What fascinated natural philosophers even more than the hallucinations they could not control were those they could. In particular, Humphry Davy's famous experiments with nitrous oxide and Thomas de Quincey's rather less

FIGURE 5.2
Punch-Drunk on Metaphysics
As always, *Punch* had a more sardonic take on the matter of scholars' hallucinations. This cartoon, titled "Metaphysics," bears the caption: "What you say about Corporeity is all very well, but it presupposes the idea of — (hic) — absolute spirituality and transcendental — (hic) — perfection — (hic); b'sides, it's incompatible — (hic) — with the def'nition of space.' — (Hic.) 'Well! — don't — go — old fellow. Have some m-m-m-ore — g-g-g-r-o-g-grog.' — (Hic.)" Though not a direct comment on the young Swiss gentleman who collapsed from six months of metaphysics, the cartoon does capture a popular sense that there was a fine line between mental and physical indulgence. Source of illustration: *Punch* 2 (1842): 149.

respectable inquiries into the effects of opium, lent credence to the thesis that some hallucinations had strictly physical causes and cures in mental and moral discipline. As part of a larger set of studies on the effects of air on health begun by Thomas Beddoes, in 1799 and 1800 Davy tested the effects of nitrous oxide on himself and his friends, including Beddoes, Samuel Taylor Coleridge, Robert Southey, James Watt, Josiah Wedgwood, Joseph Priestley, and Peter Mark Roget.[119] Two decades later, De Quincey's anonymously published *Confessions of an English Opium Eater* dramatically affirmed Davy's evidence that hallucinations could be physiologically induced.[120]

Also intriguing to natural philosophers were instances where people felt a sudden emotional pang, which they later discovered had occurred at the moment of a loved one's death. John Herschel's wife, Margaret, "had been complaining of a great depression of Spirit (very unusual with her) and in the morning (7 A.M.) expressed herself to the effect that she seemed to be looking into the dark hole of the future—as if it were a dark spot in the Sun which through the brightness allowed a glimpse of the Blackness within." Later that morning, a letter came with the news of her brother's death by "brain fever." At a Literary Society meeting eight years later, the Bishop of Oxford told Herschel of "his own 'experiences' in sympathy & clairvoyance." He had one day been seized with a powerful sense that his son, a midshipman, was in grave danger. A letter from his son a month later recalled how he had almost been killed by a falling cast iron tank. Another friend sent Herschel accounts of people who were bombarded by visions of faces and other apparitions during their waking hours. Herschel mused in his private notes,

Concerning this fantastical phenomenon, I have talked with several people, whereof some have been perfectly acquainted with it, and others have been so wholly strangers to it that they could hardly be brought to conceive or believe it. I know a Lady of excellent parts who had got past thirty without having ever had the least notice of any such thing: she was so great a stranger to it, that, when she heard me and another talking of it, could scarce forbear thinking we bantered her; but some time after drinking a large dose of dilute tea (as she was ordered by her physician) going to bed, she told us, at next meeting, that she had now experimented what our discourse had much ado to persuade her of. She had seen a great variety of faces in a long train succeeding one another as we had described; they were all strangers and intruders, such as she had no acquaintance with before, nor sought after them; and as they came of themselves, they went too; none of them staid a moment, nor could be detained by all the endeavours she could use, but went on in their solemn procession, just appeared and then vanished. This odd phenomenon seems to have a mechanical cause, & to depend upon the matter & motion of the blood or animal spirits.[121]

Herschel's use of a physical analogy ("dark spot in the Sun") in describing his wife's vision, and his skepticism of the bishop's "experiences" indicate that he was not convinced these events had any significance beyond coincidence. If he had little faith in this world's connection to the spirit realm, as we have seen he did know firsthand that such visions could be quite vivid and immune to simple mechanical explanation. Soon after dismantling his father's rotting forty-foot telescope, he one day hallucinated that it still stood in his backyard. Recovering from a fever years later, he amused himself by willing apparitions of scenes and persons.[122] During his declining years, Herschel had bustling dreams such as one containing a rapid succession of manufactures, a search for scientific apparatus, a party, and a lecture. "Oh! That pain and bustle & confusion of carts waggons scaffolding & machinery (and circular saws working horizontal & fenced off from the foot passengers) &c &c &c!!! when I expected to find all so quiet." Later that winter, during a feverish, waking night, he had a vivid visual impression of a landscape, which was lasting and rich in detail.[123] Like other natural philosophers, Herschel encouraged an experimental approach to hallucination as much to preserve his own sanity and authority as to discipline others.

Charles Babbage had his own haunted past. He shamefacedly recalled participating as a schoolboy in half-serious, half-pranking conjurings of ghosts and devils. At Cambridge, he even formed a Ghost Club. Soon after, his friend Herschel asked for any information he might have on a "Devonshire Ghost" at Chudleigh, where Babbage and his wife, Georgianna, had spent their honeymoon. The ghost, Herschel wrote, "is invisible, and . . . appears only in the shape of a noise, and I have collected many stories similar on the best authority."[124] In his adult life, Babbage felt a deep ambivalence toward his youthful interest in the supernatural, as he made clear in an extended parody of hallucinators in his autobiography. In this meditation, he described what he first thought was an encounter with a wise spirit, who told him the secrets of the universe's creation and structure. Recovering from his reverie, however, he realized that a morsel of decayed Gloucester cheese next to him in the church pew had inspired the entire vision. His mind had confused the "birth and education" of a dairy product for the development of the universe.[125]

As an insomniac adult, Babbage continued to be troubled by an active imagination. His friend Benjamin Brodie recommended putting his apparitions to work: "I frequently succeed in obtaining sleep by watching the strange, indescribable, and ever varying spectra, which I refer to the eye, though they are probably in the brain itself, and which present themselves when real objects are excluded from sight." This technique, Brodie assured, diverted attention from more complicated issues toward innocuous spectra, thus freeing the mind to rest. This seemed sound advice, coming from someone who knew the stress of a busy life: Brodie's medical practice reputedly drew £10,000 a year; he had served as serjeant surgeon to three monarchs, and was at various points president of both the Royal Society and College of Surgeons. When Babbage tried the technique, though, it backfired. He found that his powers of attention and memory were so keen that the spots transformed into faces, even whole scenes, and kept him awake most of the night. Right before his death, his father had occasional visions of women in white, and the son may have feared that he would drive himself to a similar state of mental disarray.[126]

Further examples abound of hallucinating natural philosophers in the Victorian period: Francis Galton, George Henslow, and Karl Pearson each saw spots transform into the forms of flowers and faces (see Figure 5.3).[127] One could practically open the biography of any great scholar in the nineteenth century and find that they had a hallucinatory experience. Biographers tell such stories with a mixture of delight (at the humanity of great figures) and horror (at the frailty of great figures).[128] Like the genius, the hallucinatory man of science walked the fine line between sanity and insanity.

The appropriation of genius by the Romantics made many British natural philosophers uncomfortable with the concept. The genius seemed to buy imaginative power at the expense of reason, will, and self-control. What should one make of stories such as that of the "distinguished chemist and natural philosopher" who devised experimental apparatus in his dreams? Or the Marquis de Condorcet, Benjamin Franklin, Johannes Müller, all of whom described solving puzzles and making calculations in their sleep? Exercising the mind in one's sleep might well maintain one's mental health. In fact, some saw dreaming as a God-given safety valve. If one did not sleep and dream enough, hallucinations began appearing in one's waking life. To allow the imagination too much of a foothold in the waking life was, literally, madness.[129] Thankfully, natural philosophers believed, they possessed the finest tools for regulating the frequency and interpretation of apparitions.

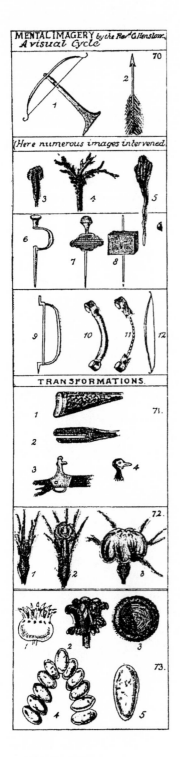

FIGURE 5.3
Henslow's "Mental Imagery"
When he shut his eyes and waited, Cambridge botanist George Henslow would usually see the "clear image of some object or other." The image would change form repeatedly, and he experimented with willing certain images to appear in succession. Once, with considerable effort, Henslow began with one image, transformed it into several others, and finally brought his visions back to the original image, creating what he called a "visual cycle." Source of illustration: Galton, *Inquiries into Human Faculty.*

Francis Galton's Study of "Mental Imagery"

As happened with color blindness, the study of hallucination eventually took a turn toward the statistical. Thanks especially to Brewster, the subject had received the kind of skeptical experimental inquiry that natural philosophers thought it deserved. Only with Francis Galton's widely read 1879–80 study of "visions of sane people," however, did the statistical prevalence of normal hallucination begin to emerge. His work also reveals the extent to which the scientific community had developed its own discourse on hallucination, one that by the late Victorian period was especially wary of spiritual connections.[130] What had begun in the early industrial period as an attempt to establish natural philosophy as a stabilizer against the threats of enthusiasm, materialism, and radical democracy, culminated in the late nineteenth century in a highly secularized and quantified new specialization of psychology.

Galton's interest in the subject probably began with his own near-hallucination during a bout with influenza and bronchitis:

When fancies gathered and I was on the borderland of delirium I was aware of the immanence of a particular hallucination. There was no vivid visualisation in it, but I felt that if I let myself go I should see in bold relief a muscular blood-stained crucified figure nailed against the wall of my bedroom opposite to my bed. What on earth made me think of this particular object I have no conception. There was nothing in it of the religious symbol, but just a prisoner freshly mauled and nailed up by a brutal Roman soldier. The interest in this to me was the severance between the state of hallucination and that of ordinary visualisation. They seemed in this case to be quite unconnected.[131]

Here, especially, we can see the effect of a century's dissolution of the ties between hallucination and faith. Galton dramatically downplayed the Christian symbolism of his vision, and instead put the episode to the service of studying the human mind. He professed no interest in whatever spiritual meaning the vision might have contained. However, in denying any spiritual meaning—how does one *almost* see Christ's figure but then not see it?—one does get the sense that Galton protested too much against the more religious sensibilities of earlier decades. But just like his forebears, Galton saw an opportunity to turn a potentially subjective experience into universal knowledge about the mind. He would later say that he took up the subject of "mental imagery," as he called it, because he thought the "essential differences between the mental operations of different men," not knowledge of the spiritual world, might provide "some clue to the origin of visions."[132]

His attention drawn to the issue of visualization and hallucination, Galton wondered about their relationship to sanity and insanity, and whether a predisposition to any of these qualities was hereditary. In November 1879, he began circulating a questionnaire on these subjects among his relatives and friends. The answers intrigued him so much that he distributed the questionnaire more widely to members of the Birmingham Philosophical Society, several boys' and girls' schools in England and the United States, and finally a sample of one hundred adult men (including nineteen fellows of the Royal Society) and one hundred adult women. Galton admitted the sampling was imperfect, but asserted that the breadth of subjects and responses far exceeded those of previous studies.[133]

His questionnaire asked the subjects to picture an object: "[S]uppose it is your breakfast-table as you sat down to it this morning." To determine the vividness of the image and the correspondence between it and the actual object, he asked several questions about illumination, definition, and coloring. He requested information about instances when the respondent mistook a mental image for reality. Finally, he probed how strong the respondent's capacity was for visualizing things such as numbers, geometric shapes, scenery, faces, machines, and military movements.[134]

Some of Galton's respondents had such developed imaging faculties, that he wondered about their sanity. Most (especially women) replied that they could often visualize landscapes, shapes, or geometrical forms—all in full color—but could only selectively employ the faculty. Galton ranked John Herschel's second-oldest daughter, Isabella, as having the highest powers of mental imagery among his sample of one hundred women. She described her vivid memory of rooms or scenery as both tactile and visual.[135]

In contrast to Isabella's skill, Galton found to his surprise that the weakest abilities in mental imagery occurred closer to home:

> To my astonishment, I found that the great majority of the men of science to whom I first applied protested that mental imagery was unknown to them, and they looked on me as fanciful and fantastic in supposing that the words "mental imagery" really expressed what I really believed everybody supposed them to mean. They had no more notion of its true character than a colour-blind man, who has not discerned his defect, has of the nature of colour. They had a mental deficiency of which they were unaware, and naturally enough supposed that those who affirmed they possessed it, were romancing.

Galton probably owed this assessment of scientists' weakness in mental imagery to his best subject, Isabella Herschel, who suggested to him that "habits of guarded and cautious expression" led men of science modestly to understate their imaging capacity. Though certainly not all such men lacked a visualizing faculty, Galton suspected that mental imagery served little use to, and might even hamper, abstract thought. Either men of science usually lost the power to visualize through disuse, or those without visualizing capability tended to gravitate to the sciences. Unwilling to simply leave this glaring limitation at the sciences' doorstep, he noted that his colleagues had other modes of thought to compensate for this lack. These different thought styles derived from the motor sense rather than the imagination, such that "men who declare themselves entirely deficient in the power of seeing mental pictures can nevertheless give life-like descriptions of what they have seen, and can otherwise express themselves as if they were gifted with a vivid visual imagination."[136] Maxwell's earlier invocation of a masculine will (see chapter 1) achieved full force here as a substitute for less trustworthy mental faculties in the well-disciplined man of science.

Among those scientific men who ranked low on Galton's scale of mental imagery capacity were the economist W. S. Jevons, physician Sir James Paget, naturalist George Romanes, astronomer Charles Piazzi Smyth, former surveyor and Toronto magnetic observatory director Henry Lefroy, and several scientific progenies such as engineers Major John Herschel and Horace Darwin. Jevons, the least visualizing of Galton's male subjects, laughed at his own frequent mistakes "in constructing a new machine so that it c[oul]d not possibly work as first made," but noted that his eyesight had no defects. A more defensive response came from the younger John Herschel, who found Galton's analogy between mental imaging and ocular vision utterly strained. "It is only as a figure of speech that I can describe my recollection of a scene as a 'mental image' which I can 'see' with my 'mind's eye.'" He had, however, seen apparitions of faces and words several times in the 1860s. John Marshall, a fellow member of the Royal Society, also wondered how Galton could place "the slightest reliance in what a *person says* of his power of seeing things with his third eye. The evidence is tainted at its very source."[137] In his defense, Galton replied that his respondents wrote their answers in "a style that testifies to much careful self-analysis," and found that subsequent "cross-examination" of his subjects "convinced me of their substantial truth."[138] Still, to his critics, none of Galton's praiseworthy statistical methods could erase the difficulty of capturing and analyzing subjective experience through personal testimony. The lingering spiritual connotations of mental imagery added to this sense that personal accounts often could not be trusted.

At the other end of the spectrum, the high number of respondents who reported not only mental imagery but hallucinations also surprised Galton. A number of people replied to his queries with stories about their apparitions of number forms, pictures triggered by words, and other figures. He knew a number of people, including three scientific men and one female relative of his, who regularly saw crowds of phantoms, or "phantasmagoria." Sir Risdon Bennett, past president of the College of Physicians, pulled Galton aside at a Royal Society meeting to describe his hallucinations. In his study at home, he sometimes saw a man in a "fantastic medieval costume" enter the room soon after he heard the postman's daily knock. Bennett enjoyed perfect health, but was still troubled by the vision, and only reluctantly made it public. Capt. J. Gore Booth of the Royal Engineers in Glasgow described to Galton feeling a sea breeze and hearing the roar of the ocean during a "fainting fit." Dr. Arthur Schuster, a fellow of the Royal Society, could call up extremely vivid images of his former instructor, James Clerk Maxwell, and could even move the phantom Maxwell in marionette fashion about imagined scenes such as the Cavendish Laboratory, his home at Glenlair, and the British Association meeting at Belfast.[139]

After studying his results, Galton again remarked on the similarity between hallucination and color blindness, since hallucinators at first thought everyone saw visions as they did. Galton suspected that, unlike the perceptions of the color-blind, hallucinations were often simple illusions compounded by the imagination. Like Maxwell, he distilled these ideas into the form of a simple equation:

$$R - S + E + H = \text{image seen}$$

where R was the actual retinal image, S the lines and shades of that image that were suppressed, E the lines and shades emphasized in the image, and H any wholly hallucinatory additions.[140]

Galton concluded that sane people not only experienced mental imagery but experienced it more than was commonly thought. Between the abstract-minded and the highly imaginative, there was a continuous spectrum of people with varying skills at mental imagery. He arranged his subjects along a "distribution of the visualizing faculty," ranking them into quartiles, octiles, and suboctiles. These abilities were also, of course, hereditary for Galton. During development, the child learned to distinguish fantasy from reality, and finally to forego fantasizing almost entirely. The shock and humiliation of hallucinating as an adult, though still perfectly natural, kept the frequency of such events from entering public knowledge. This discouragement of mental imagery character-ized periods in history when "popular opinion is of a matter-of-fact kind. . . . But let the tide of opinion change and grow favourable to supernaturalism, then the seers of visions come to the front. The faintly-perceived fantasies of ordi-nary persons become invested by the authority of reverend men with a claim to serious regard." Galton believed he had not discovered a new faculty, but had uncovered one discouraged by the Victorian tendency to naturalism.[141]

Along with the decline of supernaturalism, Galton suspected that nervous sensitivity and imagination had become so feminized and infantilized, that men did not want to admit any signs of it. It was true, he said, that women, children, the French, and "uncivilised races" such as the southern African "bushmen" tended to indulge more readily in visualization and introspection. He received support on this point from several of his subjects, including author W. Ander-son Smith, whose own imagination had diminished in the "vigour of my man-hood," and who supposed that Celts were more imaginative than Saxons.[142]

Still, for Galton the faculty seemed too valuable to abandon completely. In fact, he hoped that future generations would improve this skill through educa-tion. He had found visualizing so useful in his own anthropological work, that he commended its cultivation among other men of science.[143] To counteract a tendency to dismiss mental imagery as bogus, he noted that his trials showed these powers roughly corresponded to greater intellectual ability. In women, this might manifest itself as irritability, but men had finer gradations of sensory discrimination. In other words, men could have "a delicate power of sense dis-crimination" without the "nervous irritability" endemic to women. This was not necessarily an easy distinction to maintain, and Galton peppered his study with a schizophrenic dialogue about the desirability and undesirability of men-tal imagery. In some instances he seemed pleased with abstract thinking as a masculine alternative to nervousness. In other cases, he called these same men "lazy" for not cultivating such a useful faculty. Galton's waffling suggests once again that the masculinity of the learned man still posed some troubling para-doxes. Here, the specter of subjectivity still hovered around the imagination, continuing to foil attempts to put it to the service of masculine character and general knowledge. If a man's reason were strong enough to control the imagi-nation, however, he might profitably unleash this volatile intellectual power.[144]

CONCLUSION

In Galton's wake followed a number of studies on visions among sane persons.[145] The first major follow-up was a statistical study performed by Edmund Gurney and several other members of the Society for Psychical Research. Though his work appeared five years after Galton's, Gurney often and erroneously has received credit for offering the first statistical account of hallucination.[146] The subject of sane visions also inspired discussion at both the 1889 International Congress of Psychology meeting in Paris, and the 1892 London Congress for Experimental Psychology. The Society for Psychical Research conducted a second mammoth census with more than four hundred "collectors" and twenty-seven thousand respondents, 1,684 of whom had experienced hallucinations.[147] Related studies conducted in the United States, France, Germany, Brazil, and Russia led to one conclusion: that hallucinations could occur to otherwise normal subjects.[148]

We should expect that after undergoing its "crisis of faith," high Victorian Britain would have achieved a purely empirical approach to the problem of hallucinations.[149] If earlier natural philosophers had only *seemed* materialistic, an increasing number of scientists in the second half of the nineteenth century openly proclaimed their "scientific naturalism."[150] Again, this is not to say that spiritualism did not grip the imagination of a number of physical scientists in that period. Stewart Balfour and P. G. Tait, for instance, specifically wished to reject the separation of the natural from the miraculous. In their view, divine providence expressed itself through a transfer of energy from an invisible realm to the visible world.[151] Still, by midcentury, the notion that hallucinations lacked specific religious meaning had become commonplace for most of the scientific community. Galton found himself able to take advantage of the consensus worked out by earlier writers. He could then further generalize the subjective experience of hallucination by exploring its statistical prevalence. Far removed from the original debates about supernatural causes, he could assume the existence of natural laws governing these visions. Galton's work, then, marks the same kind of turning point for hallucination that Maxwell's and Wilson's work did for color blindness. Both placed the study of subjective visual experience within a universal context.

One thing I have sought to establish in this chapter, though, is that for much of the nineteenth century religious concerns had a more persistent grip on natural philosophy than we might expect. The motive to secularize hallucination was, ironically, partly religious. While evangelicals decried the atheistic turn of physical science, such complaints ignored the dualism that natural philosophers and physicians continued to apply to the human mind and body. The very same breed of reformers I have been discussing throughout this book—who addressed themselves, among other things, to the problem of hallucination—believed that *both* atheistic materialism *and* religious enthusiasm did a disservice to the rational organization and growth of the sciences. As John Ferriar told the Manchester Literary and Philosophical Society, "[I]t has generally been found, that an opinion, adopted without sufficient proof, is defended with an earnestness very unfriendly to investigation."[152] The incompatibility of dogmatism (whether materialist or spiritualist) and scientific investigation was a running theme in discussions of natural philosophy, and once again highlights the widespread desire to sublimate subjective experience to rational, universal laws—be they theological, scientific, or politico-economic.

Hallucinations required especially rigorous examination because of their extreme irreproducibility. Like the other visual disorders discussed in this book, only one person could see apparitions. So long as these phenomena were interpreted as supernatural communications, the personal nature of the experience posed no problem. But the plethora of information about nervous physiology, mental philosophy, and optical illusions promised a more general solution. Without denying that *some* isolated incidents of hallucination might have a divine source, early industrial natural philosophers and physicians increasingly posited naturalistic explanations. To do so required that the subjective nature of the experience be effaced in favor of generic similarities between cases. Fatigue, fever, an overactive imagination, a lack of education, and other factors were deemed common to seemingly unrelated cases of seeing apparitions. Natural philosophers knew the risk that such explanations might sound deistic or even atheistic, and typically averted that criticism by acknowledging the importance of revelation and the evidence nature gave for a Creator. By midcentury, these deflections had decreased and worn thin, but this should not retroactively diminish the earlier power of natural theology.

The new, naturalized study of hallucination encouraged not only mundane explanations for strange phenomena, but also originally served the project of rationalizing religion. To many natural philosophers and physicians, the rational study of nature constituted the surest means to both understanding God's works and keeping the peace. One reviewer wondered just after the end of the Napoleonic Wars, what would have become of people with weak nerves like us, when every church-yard was in the habit of nocturnally sending out its quota of spectres—when hobgoblins were prowling about in all directions—when you could not turn a corner but an evil-spirit came bouncing against you—when you were on no occa-sion sure of your man, who would frequently take his leave of you, without fin-ishing a sentence, in a blaze of fire—and when, with all civility be it spoken, the devil himself placed his amusement, to an extent not altogether compatible with a due sense of his personal divinity, in rambling, without any very definite object, over both town and country, and keeping a great majority of our forefathers in continual hot-water.[153]

Revelations came few and far between, and could dramatically disrupt the nerves, political order, and extensive knowledge of the natural world. A more systematic natural philosophy promised to order the vast majority of false vi-sions to establish and preserve the early industrial age's fragile stability.

Maintaining order meant explaining how so many great, seemingly sane intellectuals could hallucinate. Unlike the literature on insanity in this period, that on hallucination filled its pages not with destitute women but with well-educated people of both sexes. Women as a class could not fully achieve the bourgeois, masculine ideal of activity and rationality, but the path was certainly open to *some* women, such as Brewster's Mrs. A. and Mary Somerville. Such women—and most men—immediately recognized the need for a rational faith in place of superstition. A rational religion on a national scale promised to erase the doctrinal and philosophical differences that threatened to disrupt the concerted pursuit of truth and unity.

SIX

Conclusion

Long after scientific facts ceased to be the anomalies and exceptions Bacon
used to destroy Aristotelian axioms and natural kinds, they retained their
reputation for orneriness.

—Lorraine Daston, "Marvelous Facts"

I BEGAN THIS BOOK with an image of industrial Britain as a muscular world.
By now we know that this was also a nervous world, and not just for its women.
For men—even well-educated, well-fed men like our natural philosophers—facts
and bodies had an orneriness about them that continually defied easy control.
Creating a national science thus meant constant diligence. Natural philosophers'
labors to tame the orneriness of scientific bodies and facts fit into a broader pro-
gram designed to house persistently idiosyncratic religions, classes, ethnicities,
and other experiences under one national roof. Seeing themselves as a vital part of
the solution, natural philosophers sought to establish themselves as managers in a
new system of rational governance, one based on sound scientific reasoning.

By way of conclusion, I want to highlight and broaden the implications of
the previous chapters. Specifically, I think that we ourselves have been too cocksure that the Victorians enjoyed a cocksure confidence in their belief systems.
Other historians have shown that the Victorians actually harbored a great deal
of anxiety and uncertainty about the strength of their faith, the purity of their
sexuality, and the justice of their imperial aims.[1] I have argued that the Victorians erected their confidence in the sciences on a similarly shaky edifice. In the
century before the triumphant Great Exhibition, natural philosophers tapped at
their own crystal-palace bodies—hoping, paradoxically, to find tensile strength

in a profoundly frail nervous network. In natural philosophers' accounts of their own nervous disorder, they ingeniously cast themselves as unique adepts at the art of overcoming adversity. They claimed that their hierarchically guided empiricism not only led them through the maze of their own nervous disruption, it also could serve the national good. We have seen just how ambitious these men of science were in their desire to manage the industrial nation: those who studied color blindness claimed to have found a way to unify the British provinces into a nation; hemiopsy could demonstrate how well equipped natural philosophers were to keep the national factory at optimum efficiency; and finally, men of science proposed a moderate voluntarism as a means to peacefully unify Britain against the hallucinatory state of religious factionalism.

In the second half of the nineteenth century, science did indeed attain significantly more state power than it had ever enjoyed before in its history. Clearly, something about the sciences' self-promotion had worked. In this conclusion I will first illustrate how much that triumph relied on the natural philosopher's construction of his own nervous body by briefly examining the history of telegraphy. I will then discuss more generally how the nervous epistemology of the early industrial period can help us understand the rise of modern science in the mid-to-late Victorian period.

The Telegraph as a Scientific Nervous System

> Modern peoples by printing, gunpowder, the compass and the language of telegraph signs, have made vanish the greatest obstacles which have opposed the civilization of men, and made possible their union in great republics. It is thus that the arts and sciences serve liberty.
>
> — Bertrand Barère de Vieuzac

We might expect that the above accolade came from an English pundit, marveling at Cooke and Wheatstone's 1837 invention of the electric telegraph. In fact, however, the speaker, Bertrand Barère de Vieuzac, was a delegate to the National Convention in France, more than forty years earlier. The delegate had just witnessed the first official message passed along an *optical* telegraph line between Lille (on the border with the Austrian Netherlands) and Paris. That message reported the happy news that France was recapturing cities from the Austrians and Prussians. In 1801, Napoleon commissioned one of these optical telegraphs to facilitate an invasion of England, but later canceled the attack.[2]

Unbeknownst to Napoleon, the new technology already had found its way across the English Channel via a French prisoner of war.[3] In September 1794 the editors of the *Gentlemen's Magazine* publicized the captured intelligence to its readers: the French inventor Claude Chappe had devised the means to "transmit thoughts, in a peculiar language, from one distance to another, by means of machines, which are placed at different distances of between four and five leagues from one another, so that the expression reaches a very distant place in the space of a few minutes."[4] The article set off a host of English inventors seeking to copy or approximate this device for "communicating intelligence." With a budget of £3,000 and the design of John Gamble, chaplain to the Duke of York, the Admiralty established a line of "telegraph hills" from London to Deal by January 1796. The Admiralty continued to operate and expand the lines throughout the Napoleonic Wars,[5] using them to transmit rapidly such news as the Spithead Mutiny from ship to headquarters and back again.[6]

Designs varied between the optical telegraphs that developed, but generally they involved several movable shutters or paddles (semaphores) attached to a base. The telegraph was painted a dark color and placed on a tower, church, hill, or other high location so that its signals would contrast with the sky. Using a carefully guarded code book, a telegraph operator maneuvered the shutters or semaphores to symbolize various letters, words, or phrases. Operators read and relayed these messages along a line of stations, which typically ran between a contested borderland and a major city. The stations sat in enough proximity to each other that their operators could easily read the signals from neighboring stations using a simple telescope.[7]

Some immediately saw in the optical telegraph the potential for that Enlightenment dream: a universal language that would overcome many of provincialism's limitations.[8] Abraham Edelcrantz, a Swedish nobleman second only to Chappe in the development of the optical telegraph, hoped that his code book might eventually serve as a universal dictionary, "whereby people of all nations could communicate without knowing each other's language."[9] Semaphoric code would thus become part of the solution to the problem of communication and subjective knowledge discussed in chapter 2.

As we have already seen, however, such ideals hardly represented the realities of mechanical operation. The difficulties of working with the optical telegraph again mirrored the difficulties of working with the nervous system. Natural philosophers soon saw that in the telegraphic just as in the nervous system, communication errors frequently happened en route from source to destination. Apart from those errors caused by unavoidable external conditions, most mistakes arose from the poorly managed "automatons" who conveyed the messages. Stations thus not only had to employ an observer (usually a lieutenant) to read incoming signals and a "handyman" to reproduce those signals for the next station—but also had to keep a log of their communications so that supervisors could check them for errors.[10]

Telegraphy directors admonished their workers to correct for sensory inaccuracy by using the same mental powers of attention that were advised by the mental and moral philosophers of the period, but ultimately errors were seen as unavoidable.[11] In fact, Ignace Chappe, brother to the inventor, proclaimed it a sheer impossibility to achieve error-free transmissions down the telegraph system's long chain of stations.[12] Much of the work done to refine the optical telegraph's design during the Napoleonic Wars was directed at making the apparatus and coded language simple enough that a minimum of errors occurred, while keeping it complex enough that enemy spies could not decode the messages.[13] Because telegraphic research involved determining the minimum signaling legible with the least error over a considerable distance, it prefigured later work on just noticeable differences and later connections between the nervous system and the telegraph.[14]

By the time electric telegraphs attracted significant attention in the 1840s–'50s, the ground therefore had been laid for thinking about the analogous challenges of nervous and telegraphic communication. Since the nervous system was typically imagined as (at least like) an electrical system,[15] the analogy became especially apt with the arrival of the electric version of the telegraph. William Fothergill Cooke, one of the electric telegraph's inventors and originally an anatomical wax modeler, even referred to breaks in his wires as "injuries."[16] Herbert Spencer in his influential *Principles of Psychology* made the correspondence equally clear by comparing the evolutionary development of scientific intelligence to the expansion and workings of the global electric telegraphic network.[17] And in a particularly striking example of the nervous-telegraphic analogy, an early American historian of the telegraph, George Prescott, wrote that his device

in its most common form, communicating intelligence between distant places, performs the function of the sensitive nerves of the human body. In the fire telegraph it is made to act for the first time in its motor function, or to produce effects of power at a distance; and this is also connected with the sensitive function, through a brain or central station, which is the reservoir of electric or nervous power for the whole system. We have thus an excito-motory system, in which the intelligence and volition of the operator at the central station come in to connect sensitive and motor functions, as they would in the case of the individual. The conditions of the municipal organization absolutely compelled the relation of circuits which has been described.[18]

Prescott's vivid juxtaposition suggests a number of reasons why the analogy made sense to telegraph engineers and physiologists alike. Not only did both the telegraph and the nervous system convey intelligence and sensitivity almost instantaneously, but both also contained central stations that directed, connected, and interpreted the proceedings. This supervisory organ still proved necessary, for while the electrical telegraphic system was certainly speedier than the optical version, it still relied on mechanical and human performance, neither of which achieved perfection. Telegraphers still had to translate code into legible messages, and at an even more basic level, a messenger boy had to run that legible message to the recipient. The nonmechanical attribute known as skill—skill in both designing and working the apparatus—improved the system's efficiency.[19]

In short, natural philosophers learned through study of the nervous system and optical telegraphy that communication required supervised negotiation.[20] If signs had more than one meaning, if more than one description or perception could relate an experience, then arriving at the truth required managers to decide what would count as authoritative communication. Along with the new biblical critics, scientists and physicians configured language and even sensations as signs. Proper interpretation was no longer inherent to the signs themselves, but rather came with experience and training.[21]

By the mid-nineteenth century, it had become clear that telegraphs and nervous systems faced the same managerial problems. The intelligence that the two systems communicated was so vital that a supervisor or mind had to watch over and direct the accident-prone mechanical elements. However, the incredible success of the telegraph—both as a technological wonder and as the whip stitch of empire—also signaled the feasibility of natural philosophers' vision of a healthy, hierarchically managed knowledge network. The broad public confidence in scientific methodology that natural philosophers had struggled to achieve found a champion in the high-functioning nervous system that was the global electric telegraph.

Abnormality, the End of Natural Philosophy, and the Birth of Modern Science

After midcentury, one could detect the success of natural philosophers' hierarchical nervous epistemology not only in telegraphy, but broadly across the sciences. In fact, the institution of what we call modern science in the latter half of the nineteenth century very much depended on that success. In the stories I have told in this book about nervous epistemology we can find one answer to the question of how a relatively unified, though amorphous natural philosophy divided by the end of the nineteenth century into modern sciences such as physics, chemistry, physiology, and psychology.

Historians eagerly wish to avoid whiggism, the notion that our present state of affairs developed inexorably from a more primitive past. Avoiding this attitude is an admirable goal, and produces responsible scholarship. My concern in this section is that while we have avoided many of the pratfalls of whiggishness (for example, by studying scientific ideas that today are not considered credible), we still have some lingering problems in this area. Specifically, I am troubled that intellectual and cultural historians still primarily organize themselves according to modern disciplines, studying the history of physiology, of philosophy, of religion, of medicine, and so on. This book illuminates the extent to which a problem-focused approach can reveal aspects of the sciences that a discipline-focused approach cannot. The main reason that no one before now has noticed the prevalent nervous difficulties of such well-known men of science, is that such a topic falls between the cracks of traditional disciplinary history.

In order to truly understand how we came to have the modern scientific disciplines, we also cannot presume those disciplines have an ahistorical logic.[22] To understand how the sciences became specialized and "disciplined," we need to understand those periods when things were differently ordered. In this vein, Andrew Cunningham's point about the nature of natural philosophy is well taken. He has argued that religion was *constitutive* of Newtonian natural philosophy, not simply interactive with it. We cannot call the Newtonians' work in the eighteenth century "science," since this term connotes a very different, more distant relationship with faith. This does not mean there are no points of contact between their understanding of nature and our own. Nor is it true that natural philosophy was entirely undifferentiated from theology, for only the latter dealt in revealed knowledge. We simply must understand that modern science, by definition a secular enterprise, did not take full shape until late in the nineteenth century.[23]

Without the notion that the study of nature revealed God's design, natural philosophy necessarily ceased to exist. Historians have generally agreed that the centrality of faith in the sciences experienced a dramatic decline somewhere between the 1840s and the 1880s, and that Darwinism played only one part in this decline.[24] As I have argued, God's extrinsic relationship to nature formed a major justification for mind-body dualism and other hierarchies. Without the promise of a strong connection between natural and revealed knowledge, such hierarchies lost much of their epistemological and metaphysical foundation. That is not to say that hierarchies ceased to exist in post-natural theological science. Rather, the modern sciences forged new foundations out of particular kinds of experimentation and mathematical deduction. For example, psychology emerged as a fully independent discipline in the late nineteenth century when it considered the mind its main object of study rather than the soul.[25] Within the context of studying abnormal bodies, we have seen that a good scientific study by the latter half of the nineteenth century garnered its authority more exclusively from a secularized methodology than it had previously. As I discussed particularly in the previous chapter, the search for a rationalized, universal faith led many natural philosophers to base their conclusions on secular, even deistic arguments. While the original intent was not to damage the strength of God's presence in nature, this ultimately was the effect. In its rejection of a slavish adherence to Baconianism, the new philosophy of science put its faith in precision rather than divine fiat.[26] If God still had an important presence in Victorian science and culture more broadly, the man of science had largely usurped God's role in guaranteeing epistemological authority.

While the decline of natural theology proved vital to the emergence of the modern sciences, so did the post-Enlightenment conviction that individual subjectivities defied the familiar laws of moral philosophy and human "nature."[27] Cabanis had argued that the study of human nature could never eradicate humans' complex idiosyncrasies. Not only their thought and passions, but their bodies as well, defied all attempts at unity.[28] This conviction paralleled, and drew many tools from, a similar move toward statistical methods in the sciences.

Ultimately, though, in this context the human sciences required an identity separate from natural philosophy—and indeed natural history. After the early nineteenth century, theories of perception required a specialist knowledge of physiology and/or psychology which most of those trained in the physical sciences could not sustain. Partly because of this, the human sciences emerged as fields between 1780 and 1860.[29] For example, a separate, lasting concept of psychology (and psychiatry) developed around the turn of the nineteenth century. It grew out of the dualist sense that physiology could not reveal everything about mental life. On the other hand, as we have seen, natural philosophers and physicians also sought an empirical, experimental knowledge of thought that moral philosophy alone could not provide. The most important Enlightenment-era work on the mind in Britain came from Scots such as Whytt, Cullen, Reid, Stewart, and Brown, who believed that philosophy and physiology could provide a unified view of the mind.[30] The human mind required its own discipline, one that respected the unique and immaterial nature of the subject matter, while also using the best tools of the sciences. These tools gave one the empirical basis for avoiding speculative bogs such as the relationship between mind and matter, a problem increasingly left to the philosophers after the mid-nineteenth century. By then, through studying their own nervous abnormality, men of science had gained enough confidence in the mind-body hierarchy and in their own methodological protocols to leave the metaphysical fretting to others. The specialization and antimetaphysical inclination of modern science indicated, among other things, the belief that the sciences really could take their place in a confident industrial bureaucracy.

Finally, one of the deepest divides to form within the study of abnormal perception was between what might be called the clinical approach and the experimental approach. To understand the details of an abnormal visual experience, one had two choices: one could simply report experience in narrative fashion (the clinical approach), or one could devise instruments and other "mechanically objective" measures to reduce as far as possible the qualitative and subjective features of the observer's report (the experimental approach). Many nineteenth-century observers considered clinical medicine to be eternally—and perhaps hopelessly—tied to case histories. "Let him be ever so learned in anatomy, organic chemistry, histology, pathology, and all other sciences," wrote Samuel Brown of the physician at midcentury, "it is very seldom that he can altogether dispense with the sensations of his patient; that is to say with the reported reactions of the morbid and exceptional nerve."[31] The separation of experimental physiology and psychology from clinical medicine grew out of physiologists' and psychologists' desire to adopt the experimental approach that had proved so successful in other physical sciences.[32] In chapter 3, for example, we saw Maxwell institute the experimental approach in studies of color blindness. Where the clinical approach worked to cure individuals, men of science in the nineteenth century increasingly looked to experimental protocols to extrapolate individualized bodily cures to work for national-level problems.

All these factors—the decline of natural theology, the recognition of human subjectivity as fundamental (if in need of discipline), and the rise of experimental methods in the human sciences—contributed to the ebb of natural philosophy and the emergence of the modern sciences. I do not profess to give the last word on the birth of modern science, but this book has shown that early industrial natural philosophers' attempts to manage abnormalities (unusual bodies, figures, phenomena) give us important clues to modern science's genealogy. Specifically, we might use strategies for dealing with abnormalities to roughly mark the end of natural philosophy and the differentiation of the modern disciplines of the physical sciences, philosophy of science, experimental psychology, and physiology. Each of these fields shared an interest in the study of abnormal vision. From the jumble of strategies that natural philosophers used to study abnormal vision, we can detect a gradual hardening into different disciplines. Studies of abnormal vision alone cannot account for the splintering of natural philosophy into the specializations of modern science. But if we are ever to understand science's historical success and complex development, we must begin with cross-disciplinary views such as this, and from there advance toward a fuller understanding.

Notes

For abbreviations used to cite manuscript collections, see the beginning of the bibliography.

CHAPTER 1. THE NERVOUS MAN OF SCIENCE

1. For example, see Mathias, *The First Industrial Revolution*.
2. Showalter, *The Female Malady*. For an intelligent, though rather mean-spirited, critique of Showalter's theory that the Victorians feminized madness, see Busfield, "The Female Malady?" Showalter has since considered male nervousness in *Hystories*.
3. Gascoigne, *Cambridge in the Age of Enlightenment*, 290–91.
4. Thurston, *History of the Growth of the Steam-Engine*, chap. 3.
5. "Herschel, William," in Gillispie, *Dictionary of Scientific Biography*.
6. H. Davy, *Researches, Chemical and Philosophical*, 284.
7. Somerville, *Personal Recollections*, 183–85.
8. Babbage biography, 2, Buxton Papers, Museum for the History of Science, Oxford, cited in Hyman, *Charles Babbage*, 50.
9. Campbell and Garnett, *Life of James Clerk Maxwell*, 51, 84, 152–53, 163.
10. Spencer, *Autobiography*, 474–75, 495.
11. For example, see John Herschel to Michael Faraday, 12 March 1857, JH-RSL 23.102; David Brewster to John Herschel, 26 September 1867, JH-RSL 4.270.
12. For an excellent, related exploration of some of the points made in this chapter, see Winter, *Mesmerized*.
13. Poovey, *Making a Social Body;* Drayton, *Nature's Government;* Brewer, *Pleasures of the Imagination;* Colley, *Britons;* Vrettos, *Somatic Fictions*.
14. Hawkins, *Reconstructing Illness*. I thank Jim Capshew for this reference and several helpful discussions of pathography.
15. Cheyne, *English Malady;* Guerrini, "Case History"; Logan, *Nerves and Narratives*.
16. Logan, *Nerves and Narratives*, 2–3.
17. Cannon, *Science in Culture*, chaps. 1, 9; Heyck, "From Men of Letters to Intellectuals."
18. Cheyne, *The English Malady*.
19. A. Combe, *Observations on Mental Derangement*, 186.
20. Batten, *Resolute and Undertaking Characters*, 169–70; Otto Struve to George Airy, 14 March 1865, and Airy to Struve, 27 March 1865, RGO-CUL 6/381.699–700.

21. Claparède, *La psychologie animale de Charles Bonnet*, 21.

22. Arago and Henry entries in Gillispie, *Dictionary of Scientific Biography*; on Arago's blindness, see [Brewster], "François Arago," 475–77.

23. Purkinje and Fechner entries in Gillispie, *Dictionary of Scientific Biography*.

24. Arago, "Bailly," 1:236.

25. Helmholtz to William Thomson, 14 December 1862, WTK-CUL H70.

26. M. W. Jackson, "Artisanal Knowledge," 551.

27. Tresch, "Mechanical Romanticism."

28. Brigham, *Remarks on the Influence of Mental Cultivation*, 76–78, 119.

29. Gabbey, "Newton and Natural Philosophy." I have modified Gabbey's definition to accord with Andrew Cunningham's important points in "How the *Principia* Got Its Name." For particularities of the Scottish community, see Jacyna, *Philosophic Whigs*.

30. Bashford, *Purity and Pollution*; Kapur, *Injured Brains of Medical Minds*; Guerrini, "Case History"; Duffin, "Sick Doctors"; Altman, *Who Goes First?*.

31. J. Browne, "I Could Have Retched All Night"; Colp, *To Be an Invalid*; Secord, *Artisan Naturalists*. For analysis of a poetic case, see Cody, "Watchers upon the East."

32. Alborn, "'End of Natural Philosophy' Revisited."

33. Wise with Smith, "Work and Waste"; Cahan, *From Natural Philosophy to the Sciences*. For parallels between these changes in natural philosophy and changes in how vision was understood, see Crary, *Techniques of the Observer*.

34. As mathematician and logician Augustus de Morgan put it to Astronomer Royal George Airy, "It is not *nature* which you investigate—but matter" (De Morgan to Airy, 15 December 1855, RGO-CUL 6/376.185–86).

35. Priestley, *History and Present State of Discoveries*.

36. Clarke and Jacyna, *Nineteenth-Century Origins of Neuroscientific Concepts*, 1–28, 216; Geison, "Social and Institutional Factors."

37. Turner, "Paradigms and Productivity." Also see his *In the Eye's Mind*.

38. Jardine, "Inner History."

39. Note, though, how Brewster's daughter described the 1843 Great Schism in the Scottish Presbyterian Church, in which the Brewsters landed on the evangelical side: "Mind triumphed over matter, soul over flesh, conscience over mammon, and the gazing thousands of the city were moved into tearful admiration" (Gordon, *Home Life of Sir David Brewster*, 170–77).

40. Roger Smith first made these connections between natural philosophy and the understanding of the mind in his excellent paper, "Background of Physiological Psychology."

41. Stewart, *Elements of the Philosophy of the Human Mind*; T. Brown, *Lectures on the Philosophy of the Human Mind*. For these philosophers' profound influence on British physics, see Olson, *Scottish Philosophy and British Physics*.

42. Hartley, *Observations on Man*. I will use "mental-moral philosophy" to refer to the interrelated studies in this period of moral philosophy, philosophy of mind, philosophy of science, and nervous physiology.

43. Cullen, *Institutions of Medicine*, 1:23, 61–77, 101–102; Whytt, *Observations;* French, *Robert Whytt, the Soul, and Medicine.*

44. Elliot, *Elements of the Branches of Natural Philosophy.*

45. Bell, *Idea of a New Anatomy of the Brain;* Whytt, *Observations;* Figlio, "Theories of Perception," 178.

46. Müller, *Handbuch der Physiologie des Menschen für Vorlesungen.*

47. Holland, *Chapters on Mental Physiology*, viii, 65–66, 239–301, especially 272.

48. Whewell, *Philosophy of the Inductive Sciences*, 1:26–27, 33–37; Yeo, "Idol of the Marketplace."

49. Apart from those referred to elsewhere in this book, see also Addams, "An Account of a Peculiar Optical Phenomenon"; Griffin, "On an Unusual Affection of the Eye"; Cranmore, "On Some Phaenomena of Defective Vision"; and many others.

50. For example, see Stewart, "Some Account of a Boy Born Blind and Deaf"; Roget, "Explanation of an Optical Deception"; Scoresby, "An Inquiry into Some of the Circumstances and Principles"; and many others.

51. For example, see John Herschel to J. D. Forbes, n.d. April 1842, JH–RSL 22.115; Augustus de Morgan to George Airy, 2 September 1844, RGO–CUL 6/368.263; William Whewell to Airy, 11 August 1850, RGO–CUL 6/372.439–40; Charles Wheatstone, "Binocular Vision," CW–KCL, Section 4, Box 3, File 17, and "Colour," CW–KCL, Section 4, Box 3, File 18.

52. H. Airy, "On a Distinct Form of Transient Hemiopsia," 247.

53. For a general discussion of how the Victorians saw vision as an unstable and problematic resource, see Flint, *Victorians and the Visual Imagination.*

54. [S. Brown], "Animal Magnetism," 134.

55. Péclet, *Cours de Physique*, v; quoted in Buchwald, *Rise of the Wave Theory of Light*, 200.

56. J. Herschel, "Mathematics," 13:359; Good, "John Herschel's Optical Researches," 9.

57. J. Herschel, "Light," 4:349.

58. Young, *Mind, Brain, and Adaptation in the Nineteenth Century.*

59. One instance appears in Holland, *Chapters on Mental Physiology*, 65–68.

60. On Brown's argument see his *Lectures on the Philosophy of the Human Mind*, 1:68–69; J. Herschel, *Preliminary Discourse*, 87–88; Reed, *From Soul to Mind*, 64–72; Olson, *Scottish Philosophy and British Physics*, 29–43, 258–60.

61. Barker-Benfield, *Culture of Sensibility*, 5, 27, 91–92; Clarke, "Strenuous Idleness."

62. Maxwell, "Address to the Mathematical and Physical Sections," 2:219–20.

63. Crary, *Techniques of the Observer.*

64. Campbell and Garnett, *Life of James Clerk Maxwell*, 107–109.

65. *Autobiography of John Stuart Mill*, 118–19.

66. W. B. Carpenter, *Principles of Mental Physiology*, 25–26 (emphasis in the original).

67. Babbage, *On the Economy of Machinery and Manufactures*, 59–60.

68. Babbage, *Passages from the Life of a Philosopher*, 364–65.

69. Helmholtz, "On the Physiological Causes of Harmony in Music," 92–93.

70. For a review of this literature, see Frank, "Bringing Bodies Back In."
71. Foucault, *Birth of the Clinic;* Elias, *The Civilizing Process.* Major studies inspired by this work include Laqueur, *Making Sex;* Jones and Porter, *Reassessing Foucault;* Feher, *Fragments for a History of the Human Body.* For an excellent contextualization of Elias and Foucault, see Outram, *Body and the French Revolution,* 6–26.
72. Bordo, *Flight to Objectivity;* Keller, *Reflections on Gender and Science;* Lloyd, *Man of Reason.*
73. Haley, *Healthy Body and Victorian Culture;* R. Porter and D. Porter, *In Sickness and in Health;* Riley, *Sickness, Recovery, and Death;* Woods, "Physician, Heal Thyself."
74. Haley, *Healthy Body and Victorian Culture,* 21.
75. Rousseau and Porter, *Languages of Psyche;* Schiebinger, *Mind Has No Sex?;* Laqueur, *Making Sex;* Stafford, *Body Criticism;* Micale, *Approaching Hysteria.*
76. Morus, *Bodies/Machines;* Bud, Finn, and Trischler, *Manifesting Medicine;* Ketabgian, "Human Prosthesis"; Seltzer, *Bodies and Machines;* Adas, *Machines as the Measure of Men.*
77. Lawrence and Shapin, *Science Incarnate;* Christie, "Paracelsian Body"; Sibum, "Working Experiments"; Schaffer, "Experimenters' Techniques"; Roberts, "Death of the Sensuous Chemist"; Biagioli, "Tacit Knowledge."
78. R. Porter, *"Barely Touching,"* 49.
79. Sampson, "Establishing Embodiment in Psychology."
80. Mialet, "De-construction of a Genius," 5. I thank Ruth Benschop for telling me about this paper, and Hélène Mialet for sharing it with me.
81. Bacon, *History of Life and Death,* 125.
82. Cheyne, *Essay of Health and Long Life,* 198.
83. Ibid., 159–60 (emphasis in the original).
84. David Hume to George Cheyne?, n.d. 1734?, in Burton, *Life and Correspondence of David Hume,* 1:32; R. Porter and D. Porter, *Sickness and Health,* 210–12.
85. W. Smith, *Dissertation upon the Nerves,* 143; Trotter, *View of the Nervous Temperament,* xvii; Logan, *Nerves and Narratives,* 1–2, 19. Also see Bynum, "Nervous Patient."
86. MacKenzie, *History of Health and the Art of Preserving It,* 416–17. Carol Houlihan Flynn discusses Cheyne and MacKenzie in her "Running Out of Matter," 156–58.
87. John Herschel to Francis Baily, 25 July 1841, JH–RSL 3.212.
88. R. Porter, *"Barely Touching,"* 68–69.
89. *Specimens of the Table Talk of Samuel Taylor Coleridge,* 49.
90. Jurin, "Essay upon Distinct and Indistinct Vision"; Priestley, *History and Present State of Discoveries,* 660–63.
91. Stewart, *Lectures on Moral Philosophy . . . Delivered in the University of Edinburgh in the Years 1793–4,* 97, DS-EUL Dc.4.97.
92. Biot, *Life of Sir Isaac Newton* (this article originally appeared in the 1821 *Biographie Universelle*); Brewster, *Memoirs of the Life, Writings, and Discoveries of Sir Isaac Newton,* 2:133; Whewell, *History of the Inductive Sciences,* 2:141; Powell, *Historical View of the Progress of the Physical and Mathematical Sciences,* 534; Yeo, "Genius, Method, and Morality"; Iliffe, "Isaac Newton."

93. Wheatstone, "Contributions to the Physiology of Vision, no. I." Purkyne's own account appeared in his *Beiträge zur Kenntniß des Sehens in subjectiver Hinsicht.*

94. W. Henry, "Estimate of the Philosophical Character of Dr. Priestley," 215.

95. Arago, "Bailly," 98–100.

96. Farrar, Farrar, and Scott, "Henrys of Manchester"; Morrell and Thackray, *Gentlemen of Science*, 128–36, 402.

CHAPTER 2. THE SOCIAL HIERARCHY OF SUBJECTIVITY

1. Golinski, *Science as Public Culture*, 196–97, 284–85.

2. T. M. Porter, *Rise of Statistical Thinking*; Swijtink, "Objectification of Observation"; Schaffer, "Astronomers Mark Time"; Daston, "Objectivity and the Escape from Perspective"; Daston and Galison, "Image of Objectivity"; Rothermel, "Images of the Sun"; Wise, *Values of Precision.*

3. Boring, *Sensation and Perception*, 116–17.

4. L. Laudan, *Science and Hypothesis*, 129–31, 189; R. Laudan, "Role of Methodology in Lyell's Science"; Yeo, "Idol of the Marketplace"; Yeo, "Reviewing Herschel's *Discourse*"; Anderson, *Predicting the Weather*; Olson, *Scottish Philosophy and British Physics*, chaps. 4, 5.

5. W. Hamilton, *Discussions on Philosophy*, 243.

6. Charles Babbage to Humphry Davy, 21 May 1823, MC-RSL DM.4.130; Babbage, *Economy of Machinery and Manufactures*, 169–202, 379. Also see Hyman, *Charles Babbage*, 43–44; Romano, "Economic Ideas"; Ashworth, "Calculating Eye," especially 439–40.

7. "Arithmometer," CW-KCL Box 6.

8. John Dow, "Notes on Stewart's Lectures on Political Economy, 1808–1809," vol. 2, lecture 21, 361–62, DS-EUL Dc.3.105.

9. Roget, *Introductory Lecture*, 61–62; Jacyna, *Philosophic Whigs*, 174.

10. Anderson, "Instincts and Instruments," 155, 163–65; Green Musselman, "Worlds Displaced," 71–74.

11. Alborn, "Business of Induction."

12. Gillespie, "Divine Design and the Industrial Revolution."

13. The argument I make in this section is similar to one Christopher Lawrence has made, in which nervous physiology justified the landed minority's custodianship of civilization during the Scottish enlightenment. See his "Nervous System and Society in the Scottish Enlightenment."

14. David Ricardo to John Barton, 29 May 1817, *Letters 1816–18*, vol. 7 of *Works and Correspondence of David Ricardo*, 158–59; quoted and discussed in Berg, *Machinery Question*, 66–73, 128.

15. Babbage, *Economy of Machinery and Manufactures*, 215.

16. G. Airy, *Report of the Astronomer Royal*, quoted in his *Autobiography*, 216. For an earlier, similar sentiment, see the 1837 Board of Visitors report, quoted in Meadows, *Recent History*, 3.

17. Cited in Berg, *Machinery Question*, 158. On the goals and ultimate failure of the Mechanics' Institutes, see Inkster, "Social Context of an Educational Movement."

18. John Herschel to William Whewell, 19 August 1818, WW–TC add. ms. a.207.4. Also see Gooding, "'He Who Proves, Discovers.'"

19. Graves, *Life of Sir William Rowan Hamilton*, 1:245–46, 271–73, 285–87, 305, 431–34, 520–21. I thank the encyclopedic Simon Schaffer for this useful reference.

20. Yeo, "Scientific Method and the Rhetoric of Science," 267.

21. On the progressive conservatism of Herschel and Brewster, see Alborn, "'End of Natural Philosophy' Revisited," 249. Also, see Herschel's attack against whigs and radical democrats in his unpublished "Scraps of Philosophy, of Morals, Politics, Physics, Pol[itical] Oecon[omy] &c," n.d., JH–HRC, W0065.

22. Barton, "'Men of Science,'" 108.

23. Secord, *Artisan Naturalists*.

24. Holland, *Chapters on Mental Physiology*, 1.

25. Recent work on local versus national governance includes Eastwood, *Governing Rural England;* Langford, *Public Life and the Propertied Englishman;* Prest, *Liberty and Locality*. On the rise of reform-minded science, see Miller, "Between Hostile Camps"; Desmond, *Politics of Evolution;* and sources cited elsewhere in this section.

26. Wahrman, "National Society, Communal Culture," 45. Also see Wahrman, *Imagining the Middle Class*.

27. Wahrman, "National Society, Communal Culture," 71–72; Thackray, "Natural Knowledge."

28. J. Herschel, *Preliminary Discourse*, 133.

29. Inkster and Morrell, *Metropolis and Province*. For much of the background information presented here on provincialism and science, I am indebted to this volume and Morrell and Thackray, *Gentlemen of Science*.

30. Gascoigne, *Joseph Banks and the English Enlightenment*, chap. 6; Miller, "Royal Society of London"; Gleason, *Royal Society of London*.

31. Morrell, "Economic and Ornamental Geology." Also see Orange, *Philosophers and Provincials*.

32. Sorrenson, "Towards a History of the Royal Society in the 18th Century"; Roberts, "Death of the Sensuous Chemist," 514–19.

33. Cawood, "Terrestrial Magnetism"; Cawood, "Magnetic Crusade"; Morrell and Thackray, *Gentlemen of Science*, 353–70; Stafford, "Geological Surveys."

34. Hamlin, *Public Health and Social Justice*.

35. Humphry Davy, unspecified fragment of a ms. on genius, quoted in J. Davy, *Memoirs of the Life of Sir Humphry Davy*, 1:114–15.

36. Morrell, "Individualism and the Structure of British Science in 1830." Also see Hacking, *Taming of Chance;* Merz, *History of European Thought in the Nineteenth Century*, 1:229–301.

37. Orange, "British Association for the Advancement of Science"; Morrell and Thackray, *Gentlemen of Science*, 92–93, 124–25; Lowe, "British Association and the Provincial Public"; Yeo, "Scientific Method and the Image of Science." On a similar, though earlier movement in medicine, see Warner, "Idea of Science in English Medicine."

38. Babbage, *Economy of Machinery and Manufactures*, 384. In couching—a procedure still used today to treat cataract—the lens of the eye is displaced into the vitreous humor. Several years after Babbage's book appeared the Duke of Sussex had his cataracts couched. This is ironic, since the duke had defeated the reformers' man of long-range vision, John Herschel, in the controversial 1830 election for the presidency of the Royal Society (Francis Baily to John Herschel, 2 May and 16 August 1836, JH–RSL 3.133 and 3.135).

39. Pickstone, "Bodies, Fields, and Factories"; Miller, "Royal Society of London."

40. Thackray, "Natural Knowledge."

41. Ibid., 684. On the Unitarians' importance to Newcastle science, see Orange, "Rational Dissent and Provincial Science."

42. RMI Council Minute Book, 445–46, M 6/1/1/2, Manchester Central Reference Library; quoted in Wach, "Culture and the Middle Classes," 388.

43. Neve, "Science in a Commercial City"; Kitteringham, "Science in Provincial Society." Also see Smith and Wise, *Energy and Empire*, especially chap. 2.

44. Emerson, "Philosophical Society of Edinburgh"; Emerson, "Scottish Enlightenment"; Emerson, "Science and the Origins and Concerns of the Scottish Enlightenment"; Shapin, "Audience for Science"; Morrell, "Reflections on the History of Scottish Science."

45. Shapin, "'Nibbling at the Teats of Science,'" 170–71.

46. Schaffer, "Physics Laboratories and the Victorian Country House."

47. Barker-Benfield, *Culture of Sensibility*.

48. S. T. Coleridge, *Aids to Reflection*, 27.

49. Stewart, *Lectures on Moral Philosophy*, 197; Danahay, *Community of One*.

50. This phrase began a short poem that appeared on the cover of each issue.

51. Quoted in [Brewster], "Memoirs of John Dalton," 471.

52. *Autobiography of Charles Darwin*, 139, 140–45.

53. Faraday, "Observations," 207–211, 213–16, 220–22.

54. Ibid., 206–207 (emphasis mine).

55. Abercrombie, *Inquiries*, 283. Also see Brodie, *Psychological Inquiries*, 246–47.

56. For example, see Tuana, *Less Noble Sex*, 97–101; Moscucci, *Science of Woman*.

57. For example, see Mullan, "Gendered Knowledge, Gendered Minds," 41–56; Winter, "Calculus of Suffering."

58. George Airy to unnamed recipient, 10 November 1880, quoted in his *Autobiography*, 332–33 (emphasis in the original).

59. Bacon, *Masculine Birth of Time*, 1602/3; see Benjamin Farrington's translation in "*Temporis Partus Masculus*"; discussed in Keller, *Reflections on Gender and Science*, 38.

60. Stafford, *Voyage into Substance*, 474. Also see Keller, *Reflections on Gender and Science*, 39–42.

61. [J. Herschel], "*Mechanism of the Heavens*," 551.

62. Mary Somerville to John Herschel, 28 May 1831, JH–RSL 16.344.

63. [J. Herschel], "*Mechanism of the Heavens*," 537.

64. Ibid., 548.

65. Ibid.

66. Jane Marcet to Mary Somerville, 6 April 1834, quoted in Somerville, *Personal Recollections*, 209-10. For Martha Somerville's assessment see ibid., 6.

67. Somerville, *Personal Recollections*, 46.

68. Wollstonecraft, *Vindication of the Rights of Women*, 105.

69. James Clerk Maxwell to Lewis Campbell, 10 February 1852, in Campbell and Garnett, *Life of James Clerk Maxwell*, 176-77.

70. Caroline Herschel to John Herschel, 5 December 1826 and 21 August 1828, CH-BL Egerton ms. 3761; Caroline Herschel to Nevil Maskelyne, 3 March 1794, CB-BL add. ms. 37203.6. Also see Ogilvie, "Caroline Herschel's Contributions to Astronomy."

71. Mitchell, "Maria Mitchell's Reminiscences of the Herschels," 908.

72. Ashworth, "Memory, Efficiency, and Symbolic Analysis." Also see a fascinating Babbage document on the formation of a language for mechanical drawing, CB-BL add. ms. 37204.1-61.

73. Knox, "Dephlogisticating the Bible." Knox makes a very interesting comparison between the work of Cambridge's Bible Society, which sought to distribute the Bible without the Anglican Church's prayer book, and that of the Analytical Society, which sought to disseminate an unadulterated French mathematics.

74. Carbutt, "Essay on the Signs of Ideas"; Swift, *Gulliver's Travels*, 192-93. Gulliver marveled at two Laputans who attempted to converse nonverbally by showing each other objects which they carried with them in enormous backpacks.

75. Daston, "Knowledge of the Invisible."

76. Hannaway, *Chemists and the Word*, 146.

77. Gooding, "How Do Scientists Reach Agreement about Novel Observations?"

78. Locke, *Essay concerning Human Understanding*, 397-401; Sprat, *History of the Royal-Society of London;* Cantor, "Light and Enlightenment," 73; Shapin, *Social History of Truth*.

79. Aarsleff, *Study of Language in England;* Helmholtz, "Recent Progress of the Theory of Vision," 274.

80. Wartmann, "Memoir on Daltonism," 4:164. This paper was originally published as "Mémoire sur le Daltonisme," *Mémoires Soc. Phys. Genève* 10 (1843): 274-326.

81. [Brewster], "Dr. George Wilson *On Colour-Blindness*"; Shapin, *Social History of Truth*.

82. Helmholtz, "On Goethe's Scientific Researches," in *Popular Lectures*, 54.

CHAPTER 3. PROVINCIALISM AND COLOR BLINDNESS

1. Wordsworth, "The Excursion," 5:247 (book 7, lines 507-15). Wordsworth met Gough on one of his visits to Thomas Cookson. See Reed, *Wordsworth*, 324.

2. [Brewster], "Memoirs of John Dalton," 472; Nicholson, *Annals of Kendal*, 355-68 ; "Gough, John," in Stephen, *Dictionary of National Biography*.

3. John Dalton to Peter Crosthwaite, 12 April 1788, quoted in W. C. Henry, *Memoirs of the Life*, 9-10.

4. Gough also tutored two other Cambridge wranglers: Joshua King, later master of Queen's College, and Thomas Gaskins, a tutor of Jesus College. See Patterson, *John Dalton and the Atomic Theory*, 27, 67. Gough was not the only well-known, blind, provincial figure. The blind chemist Henry Moyes was one of the most popular itinerant lectures in the late eighteenth century. He supposedly could smell black clothing. Also blind were Nicholas Saunderson, Lucasian professor at Cambridge from 1711 to 1739; John Metcalf, a road and bridge engineer whose mathematical prowess Samuel Smiles celebrated; and John Speke, the first European to reach the source of the Nile. See Harrison, "'Blind Henry Moyes'"; Wilson, *Biography of the Blind;* Tattersall, "Nicholas Saunderson"; Smiles, *Lives of the Engineers,* 1:208–34; Abercrombie, *Inquiries concerning the Intellectual Powers,* 45–46.

5. British microscopist Jabez Hogg claimed that Erasmus Darwin also was color-blind, but I can find no evidence to corroborate this. See Hogg's "Colour-Blindness," 502.

6. This passage appears in both Babbage, *Exposition of 1851,* viii–xi; and Babbage, *Passages from the Life of a Philosopher,* 128.

7. Quoted in Tyndall, "On a Peculiar Case of Colour Blindness," 333.

8. Burney, *Madame D'Arblay's Diary and Letters,* 1:307.

9. Colquhoun, "Account of Two Cases."

10. Wilson, "Dalton," 559.

11. McDermot, *Critical Dissertation,* 5–10.

12. W. E. Gladstone, "Colour-Sense," 366–68; W. E. Gladstone, *Studies on Homer,* 3:457.

13. W. Henry, "Estimate of the Philosophical Character of Dr. Priestley," 214.

14. Wartmann, "Memoir on Daltonism," 171.

15. The controversy hinged on Jamaica governor Edward John Eyre's execution of 439 people in response to an 1865 rebellion by several hundred blacks in Morant Bay.

16. "Last Case of Colour-Blindness," 216.

17. Wilkinson, *On Colour;* [Brewster], "Form and Colour."

18. [Brewster], "Form and Colour."

19. Babbage, *Exposition of 1851,* 123–26; Barnes, "Fashioning a Natural Self."

20. Kuklick, "Color Blue."

21. Darwin, *Autobiography,* 91.

22. Secord, *Artisan Naturalists;* Hermann, *History of Astronomy;* Shapiro, *Fits, Passions, and Paroxysms;* Hakfoort, *Optics in the Age of Euler,* 5, 11–26; McGucken, *Nineteenth Century Spectroscopy;* Roberts, "Death of the Sensuous Chemist."

23. Newton, *Opticks,* 150–54. These pages correspond to book I, part II, proposition V, theorem IV, experiment 15.

24. For example, see Chambers, *Handbook of Descriptive and Practical Astronomy,* 2:278–79; Smyth, *Cycle of Celestial Objects,* 1:302; Wilson, *Researches on Colour-Blindness,* 86.

25. Dickens, *Dombey and Son,* 37; Horton, "Were They Having Fun Yet?," 1.

26. McGuire and Tamny, *Certain Philosophical Questions,* 241–74; Duck, "Newton and Goethe on Colour," 507–11.

27. Sepper, *Goethe contra Newton*, x, 6–19, 27–38, and *passim*.
28. Newton, *Opticks*, 82–107, especially 102–103. These pages correspond to book I, part I, proposition VII.
29. Hutchison, "Idiosyncrasy, Achromatic Lenses, and Early Romanticism." Also see M. W. Jackson, "A Spectrum of Belief."
30. M. W. Jackson, "Artisanal Knowledge."
31. J. Herschel, *Preliminary Discourse*, 246–47.
32. Yeo, *Defining Science*, 95–96.
33. Shapiro, *Fits, Passions, and Paroxysms*.
34. General histories of color blindness research ca. 1780–1860 can be found in Wartmann, "Memoir on Daltonism" and Sherman, *Colour Vision in the Nineteenth Century*, chap. 7.
35. Brewster, "Notice Respecting Certain Changes of Colour"; [Brewster], "Dr. George Wilson," 343–46.
36. G. Combe, *Elements of Phrenology*, 97–99; G. Combe, *System of Phrenology*, 2:481–98.
37. Young, *Course of Lectures*, 2:315–16. In his original 1801 theory, Young had argued more traditionally that red, yellow, and blue were the primaries. See his "On the Mechanism of the Eye"; Hargreave, "Thomas Young's Theory of Color Vision," 91–185. This dissertation—unfortunately, never published as a book—provides an insightful overview of physiological optics in early-nineteenth-century Britain.
38. [Brewster], "Dr. George Wilson," 346; Brewster, "Observations on Colour Blindness," 135. Brewster elsewhere described physiological optics—something to which he devoted a great deal of time—as the most speculative, and therefore most regressive, of the physical sciences. See his *Memoirs of the Life, Writings, and Discoveries of Sir Isaac Newton*, 1:231.
39. Hargreave, "Thomas Young's Theory of Color Vision," 79–83.
40. J. Herschel, "Light," 434.
41. Figlio, "Theories of Perception," 179–85.
42. [Brewster], "Dr. George Wilson," 347.
43. *Werner's Nomenclature of Colours*; Chevreul, *De la loi du contraste simultané des couleurs*; Hay, *Nomenclature of Colours*.
44. Thomas Forster, for example, expressed his wish for such standard measures in "On a Systematic Arrangement of Colours."
45. Wartmann, "Memoir on Daltonism," 176–78.
46. Newton, *Opticks*, 125–29. These pages correspond to book I, part II, proposition III, problem I, experiment 7.
47. The literature on the subject of methodological changes from case narratives to standardized testing is vast, but see Fissell, "Disappearance of the Patient's Narrative"; Daston, "Objectivity and the Escape from Perspective"; T. M. Porter, "Objectivity as Standardization"; Matthews, *Quantification and the Quest for Medical Certainty*. Kathryn Montgomery Hunter notes the persistence and necessity of case histories in clinical medicine in her *Doctor's Stories*.
48. Note, for instance, the subject's absence from several prominent, synthetic works from the turn of the nineteenth century: Magendie, *Précis Elémentaire de Physio-*

logie; Müller, *Zur vergleichenden Physiologie des Gesichtssinnes;* Priestley, *History and Present State of Discoveries.*

49. Tubervile, "Two Letters."

50. De La Hire, "Dissertation sur les differens accidens de la vue"; Boyle, "Some Uncommon Observations."

51. Huddart, "Account of Persons Who Could Not Distinguish Colours." The man may have been, or been related to, one Thomas Harris, listed as a Manchester shoe merchant in *Bancks's Manchester and Salford Directory,* 20.

52. Scott, "Account of a Remarkable Imperfection of Sight."

53. Veitch, "Memoir of Dugald Stewart," 10:clxiv-clxv.

54. Stewart, *Elements of the Philosophy of the Human Mind,* 57; Tannoch-Bland, "Dugald Stewart on Intellectual Character."

55. Dalton, "Extraordinary Facts." He read the paper on 31 October 1794. Dalton's paper was reprinted in the *Edinburgh Journal of Science* 5 (1831): 88-98. For another firsthand account of Dalton's discovery, see his letter to Elihu Robinson, 20 February 1794, reprinted in Lonsdale, *John Dalton,* 101-12. Dalton's discovery narrative was later recounted in Roscoe, *John Dalton;* Millington, *John Dalton;* Wright, "Unsolved Problem of 'Daltonism.'"

56. This comment came from a former pupil, Mrs. Cookson, quoted in W. C. Henry, *Memoirs of the Life,* 5.

57. Thomas Greenup to John Dalton, 9-10 April 1790, quoted in W. C. Henry, *Memoirs of the Life,* 15; Lonsdale, *John Dalton,* 74-77.

58. Lonsdale, *John Dalton,* 19-20. In a review of Henry's biography, George Wilson devoted nearly as much space to Dalton's early frustrated circumstances and character quirks as to his chemical philosophy. See Wilson, "Dalton."

59. Thackray, *John Dalton,* 42-48, 142, 144-46; Patterson, *John Dalton,* 44-46.

60. Thackray, *John Dalton,* 23, 42-48, 54-55, 85.

61. Dalton lecture in Edinburgh in 1807, quoted in Roscoe, *John Dalton,* 166-67.

62. Geikie, *Life of Sir Roderick I. Murchison,* 1:187-88 (emphasis in the original); Morrell and Thackray, *Gentlemen of Science,* 97.

63. "Members of the Manchester Meeting 3 mo. 1794," JD-JRL 144. The majority of Dalton's manuscripts were destroyed in a bombing raid on Manchester in 1940. During a series of 1803 lectures at the Royal Institution, Dalton performed the following stunt while discussing optics: "I got six ribbands, blue, pink, lilac, and red, green and brown, which matched very well, and told the curious audience so. I do not know whether they generally believed me to be serious, but one gentleman came up immediately after, and told me he perfectly agreed with me; he had not remarked the difference by candle-light" (John Dalton to Jonathan Dalton, 1 February 1804, quoted in W. C. Henry, *Memoirs of the Life,* 47-49). Also see Dalton's lecture notes, "Optics Lectures," JD-JRL 84.

64. Dalton, "Extraordinary Facts," 30-31, 37-40; John Dalton to Elihu Robinson, 20 February 1794, in Lonsdale, *John Dalton,* 101-2. Also see his test of an unnamed boy and girl in "Optics Lectures," JD-JRL 84.

65. Dalton, "Extraordinary Facts," 41-43.

66. Joseph Ransome to George Wilson, n.d., quoted in W. C. Henry, *Memoirs of the Life*, 198–203; [Brewster], "Memoirs of John Dalton," 477; Duveen and Klickstein, "John Dalton's Autopsy"; Hunt, Dulai, Bowmaker, and Mollon, "Chemistry of John Dalton's Color Blindness"; Sherman, *Colour Vision*, 127–28.

67. Young was born a Friend, but later lived as an Anglican. See Peterfreund, "Scientific Models in Optics," 69–70.

68. Huddart, "Account of Persons Who Could Not Distinguish Colours," 261–62.

69. The following letters detail Dalton's instructions and the Harrises' responses: John Dalton to Joseph Dickinson, 3 October 1793?; Dickinson to Dalton, 8 February 1794; Dalton to Dickinson, 13 April 1794; Dickinson to Dalton, n.d.; all reprinted in Lonsdale, *John Dalton*, 102–9. Also see Sherman, *Colour Vision*, 121–26. As fate would have it, Dickinson's son Isaac later turned out to be color-blind, but objected to the label, considering his color sense different rather than deficient (see Lonsdale, *John Dalton*, footnote on 109–10).

70. Reported in [Brewster], "Dr. George Wilson," 335–36.

71. Patterson, *John Dalton*, 6; Lonsdale, *John Dalton*, 29–31.

72. Quoted in Pointon, "Quakerism and Visual Culture," 407.

73. Clarkson, *Portraiture of Quakerism*, 15–26, 103–15.

74. Ibid., 119–40.

75. Southey, "History of Dissenters," 111–12; Clarkson, *Portraiture of Quakerism*, 139.

76. Russell, *History of Quakerism*, 225, 229–30, 251–98, 331–41.

77. Dalton, *Elements of English Grammar*. In April 1798 at the Lit and Phil, he read a paper that was never published, "Essay on the Mind, Its Ideas, and Affections; with an Application of Principles to Explain the Economy of Language." See W. C. Henry, *Memoirs of the Life*, 208–9; Patterson, *John Dalton*, 43, 75–78, 157–58, 188–90.

78. Roscoe, *John Dalton*, 70–71; Lonsdale, *John Dalton*, 99–100.

79. [Brewster], "Memoirs of John Dalton," 490–92. John Henry Newman was likewise horrified by the British Association's conferral of DCL degrees on four dissenters in the University church, where Newman was vicar. See Morrell and Thackray, *Gentlemen of Science*, 232.

80. Charles Babbage to William Charles Henry, 7 February 1854, quoted in W. C. Henry, *Memoirs of the Life*, 185–89. The full letter can also be found in Babbage, *Passages from the Life of a Philosopher*, 219–22. Though more experienced at court, Babbage must have empathized with Dalton. When Babbage had visited Trinity College, Dublin, during the 1835 British Association meeting, he inadvertently offended the Protestant fellows with his green waistcoat. The local tailor only had a multicolored cloth to offer as an alternative, so Babbage was left to earn the reputation as a dandy because of his wish to remain politically neutral. He believed that such neutrality was the only way to keep the fraught Irish provinces content with a national framework. (See Hyman, *Charles Babbage*, 174.)

81. W. C. Henry, *Memoirs of the Life*, 3; Wilson, "Dalton," 555, 561; Lonsdale, *John Dalton*, 50–52.

82. Raistrick, *Quakers in Science and Industry*, 271.

83. William Henry to Charles Babbage, 19 May 1830, CB-BL add. ms. 37185.179–80; Charles Babbage to Thomas Spring Rice (Joint Secretary of the Treasury), 7 Au-

gust 1831, CB-BL add. ms. 37186.37. For Dalton's agreement with Babbage's views in *Reflections on the Decline of Science in England*, see John Dalton to Charles Babbage, 15 May 1830, CB-BL add. ms. 37185.176. Also see Dalton to Babbage, 7 December 1830, CB-BL add. ms. 37185.370.

84. William Henry to Charles Babbage, 22 February? 1832, CB-BL add. ms. 37186.260-63; W. C. Henry, *Memoirs of the Life*, 173-81; Patterson, *John Dalton*, 231-35; Morrell and Thackray, *Gentlemen of Science*, 326-28.

85. [Holland], "Life of Dalton," 49.

86. Humphry Davy, sketch of Dalton, dated February 1829, quoted in W. C. Henry, *Memoirs of the Life*, 126-27; Davy, "Presidential Address," 92-99; Patterson, *John Dalton*, 217-21; Miller, "Between Hostile Camps."

87. Wilson, *Religio Chemici*, 310; Wilson, *Researches on Colour-Blindness*, viii-ix. Galton uncritically cited this "fact" several decades later in his *Inquiries into Human Faculty*, 40.

88. "Blunders of Vision," 517.

89. Ibid., 513.

90. Guerrini, "Case History"; Appleby, "Human Curiosities."

91. Sherman, *Colour Vision*, 136. Papers that briefly mentioned color blindness included Palmer, *Theory of Colors and Vision*; Goethe, *Zur Naturwissenschaft überhaupt*, 1:297; Gall and Spurzheim, *Anatomie et physiologie du système nerveux*, 4:98; Wardrop, *Essays on the Morbid Anatomy of the Human Eye*, 2:196; Rozier, "Observations sur quelque personnes," 86-88; Helling, *Praktisches Handbuch der Augenkrankheiten*, 1:1; Butter, "Remarks on the Insensibility of the Eye." For important, in-depth, continental studies, see Goethe, *Zur Farbenlehre*, 354-59; Sommer, "Über Chromatopseudopsie"; Seebeck, "Über den bei manchen Personen vorkommenden Mangel an Farbensinn"; Szokalski, "Essai sur les sensations des couleurs"; Hombres-Firmas, "Observations d'achromatopsie"; Hombres-Firmas, "Nouvelles observations d'achromatopsie."

92. Forbes, *Danger of Superficial Knowledge*, vii; discussed in Yeo, "Scientific Method," 75-78.

93. Coon, "Standardizing the Subject." Coon describes American psychology, which committed itself even more thoroughly to eradicating subjective methods than did British students of the mind, who—I am arguing—sought instead to *manage* subjectivity.

94. Alborn, "Wasted Work"; Fissell, "Disappearance of the Patient's Narrative"; Foucault, *Birth of the Clinic*.

95. Marey, *La Méthode graphique*, iii-vi; Daston and Galison, "Image of Objectivity," 81-128.

96. Turner, "Paradigms and Productivity."

97. Harvey, "On an Anomalous Case of Vision," 253, 260.

98. McConnell, *Instrument Makers to the World*, 8. Also see Troughton, "Account of a Method of Dividing."

99. McConnell, *Instrument Makers to the World*, 12.

100. Sheepshanks, "Memoir of Edward Troughton," 288 (emphasis in the original). On Troughton's visual acuity, see 284, 286.

101. Ibid., 283; George Airy to Isaac Fletcher, 20 September 1866, RGO–CUL 6/381.319–20.
102. J. Herschel, "Light," 434–35.
103. John Herschel to David Brewster, 31 January 1824, JH–RSL 20.174–75. In this letter, Herschel mentions his study of Troughton and his intent to publish it in the future.
104. J. Herschel, "Light," 435. Though this article was not published until 1845, it was signed by Herschel on 12 December 1827, in circulation by 1828, and available in French and German translation in the early 1830s.
105. Herschel first recorded this conclusion in a letter to Dalton, 20 May 1833, JH–RSL 25.3.15. This letter is reproduced in W. C. Henry, *Memoirs of the Life*, 25–27.
106. William Herschel's interest in vision centered, of course, on his telescope construction and attempts to see farther into space. See, for instance, his "Investigation of the Cause."
107. John Herschel, "Queries sent to Mr. May to be forwarded to individuals who have an imperfect perception of colour," JH–HRC M-W0072. This manuscript is not dated, but one of the responses is dated September 1832, and it seems reasonable to assume that Herschel drafted the instructions earlier that year.
108. It is unclear exactly how many people Herschel tested. Documents remain for tests on four people (not including Herschel, who undoubtedly used himself as a normal-sighted comparison), but he claimed to have known a "Dozen now living within the circle of my own enquiry." See his letter to an unknown recipient, n.d., labeled "The Eye," JH–HRC M-W0067.
109. John Herschel to Charles May, n.d. (but probably mid-November 1832), JH–HRC L0254. Herschel sent a version of this letter along with a copy of the "Queries" cited above. Charles May to John Herschel, 23 November 1832, JH–HRC M0351; Charles May to John Herschel, 7 March 1833, JH–RSL 12.310.
110. "Joseph Foster, Bromley, Middl^x, age ab^t 70," JH–HRC M1154; "John Dalton, answers to Sir J. F. W. Herschell's [*sic*] enquiries," JH–HRC M1187. Some of Dalton's replies also appeared much later in Pole, "Daltonism."
111. "John Dalton, answers," JH–HRC M1187.
112. "Mrs. Malthus's answers to the general queries," JH–HRC M1142. For the bulk of Malthus's report, see William Whewell to John Herschel, 26 October 1832, JH–HRC M0565. Though they have been cataloged separately, I believe them to have been originally parts of the same document.
113. William Whewell to John Herschel, 14 September 1832, JH–HRC M0564. On Herschel's continued interest in color blindness, see Herschel diary, 2 and 7 December 1836, JH–RSL MS 583.
114. Wilson, *Researches on Colour-Blindness;* but also see his earlier work: "On the Prevalence of Chromato-pseudopsis."
115. For biographical information, see Wilson, *Memoir of George Wilson*.
116. Wilson, *Researches*, vi–vii, 14–40. In fact, one of his students, William Bryce, probably sparked Wilson's initial interest in color blindness. Wilson realized this prize medal-winning medical student's color blindness after becoming frustrated with teaching him experimental chemistry.

117. Ibid., 10.

118. Moigno, *Repertoire d'optique moderne.*

119. Tyndall, "Coloured Lamps on Railways," 136–37; Wilson, "On Colour-Blindness," 418.

120. [Brewster], "Dr. George Wilson," 327.

121. Wilson, *Researches,* 123–25.

122. [Brewster], "Dr. George Wilson," 349–56.

123. Turner, *In the Eye's Mind,* 177–78.

124. Burch, "On Artificial Temporary Colour-Blindness"; Jeffries, *Color-Blindness.*

125. "Report of the Committee on Colour-Vision."

126. Thomas Herbert Bickerton, "Notes on Colour-Blindness, Liverpool Royal Infirmary, 1885–1901," WLHUM MS.1162; "Report of the Committee on Colour-Vision," 314–28.

127. Pole, "Daltonism," 835–36.

128. Helmholtz was long credited with revitalizing Young's theory, but it has been well established that he rejected the theory until he read Maxwell's work. See, for instance, Hargreave, "Thomas Young's Theory," 342–43; Lenoir, "Helmholtz and the Materialities of Communication"; Sherman, *Colour Vision,* 217–21. On Wilson's prompting of Maxwell, see James Clerk Maxwell to William Thomson, 13 September 1855, WTK-CUL M96.

129. James Clerk Maxwell to William Thomson, 14 March 1854, WTK-CUL M88.

130. Maxwell, "On the Theory of Colours," reprinted in Niven, *Scientific Papers,* 1:124.

131. Campbell and Garnett, *Life of James Clerk Maxwell,* 5–12, 162, 368.

132. Ibid., 12–13, 35–36, 52.

133. Hay, *Principles of Beauty,* 7–8.

134. Hay, *Nomenclature of Colours;* Everitt, *James Clerk Maxwell,* 65–66. William Thomson ordered his own set of papers from Hay in 1856, though it is unclear whether or how he used them. See William Thomson to David Ramsay Hay, 29 January 1856, DRH-EUL Dc.2.59.

135. Maxwell, "Experiments on Colour," reprinted in Niven, *Scientific Papers,* 1:126–54.

136. Ibid., 136–42; Sherman, *Colour Vision,* 175–80. In 1861, Maxwell demonstrated the trichromatic photograph experiment at the Royal Institution (Everitt, *James Clerk Maxwell,* 71–72).

137. Maxwell, "Experiments on Colour," 138; Sherman, *Colour Vision,* 189–97.

138. James Clerk Maxwell to G. G. Stokes, 23 April 1860, GGS-CUL M416.

139. On the issue of how much Katherine contributed to James' research, see Green Musselman, "Katherine Clerk Maxwell."

140. For example, see James Clerk Maxwell to G. G. Stokes, 7 May 1860, GGS-CUL M417.

141. James Clerk Maxwell to G. G. Stokes, 8 October 1859, GGS-CUL M411.

142. For a description of their experiments, see Sherman, *Colour Vision,* 197–203, 211–17, and Maxwell's letters to G. G. Stokes, 8 October 1859 to 10 September 1862, GGS-CUL M411–19. These letters are reprinted in *Memoir and Scientific Correspondence of the Late Sir George Gabriel Stokes,* 2:14–22. On Wundt, see Danziger, *Constructing the Subject,* especially 34–39; Kusch, *Psychological Knowledge.*

143. Maxwell, "On the Theory of Compound Colours," reprinted in Niven, *Scientific Papers*, 1:420–22; Maxwell, "Description of an Instrument," reprinted in Harman, *Scientific Letters and Papers*, 1:600–2; Maxwell to G. G. Stokes, 24 December 1859, GGS-CUL M412; and 25 February 1860, GGS-CUL M414.

144. James Clerk Maxwell to G. G. Stokes, 25 February 1860, GGS-CUL M414. Maxwell likely had also used several of these subjects for his 1855 paper.

145. Sherman, *Colour Vision*, 169–70.

146. Herschel noted in his 22 August 1857 diary entry: "All day Optical Experiments with Mr Pole whose vision is is [*sic*] assuredly Dichronic & whose eyes receive sensation of light from all the spectrum." Herschel diary, JH-RSL MS 585. He wrote de Morgan on 11 November 1856 that "I am about Colour blindness; will you (at your leisure) tell me how many persons *in the number of your speaking acquaintance* cannot tell scarlet from green?" (JH-RSL 23.183). De Morgan does not seem to have replied.

147. Pole, "On Colour-Blindness," 324–25.

148. Charles Wheatstone, 4 August 1856, RR-RSL 3.215.

149. G. G. Stokes to William Pole, 22 August 1859, GGS-CUL P413; Pole to Stokes, 5 October 1859, GGS-CUL P409.

150. John Herschel to John Dalton, 20 May 1833, JH-RSL 25.3.15. Also see George Airy to John Herschel, 26 August 1859, JH-RSL 1.242. For Brewster's increasingly unpopular view on color, see Cantor, "Brewster on the Nature of Light"; Brewster, *Letters on Natural Magic*, 29–32. On Helmholtz and color mixing, see Hargreave, "Thomas Young's Theory," 117–24, 271–302.

151. JH-RSL MS 625.

152. Pole, "On Colour-Blindness," 323; Pole, "On the Present State of Knowledge"; Pole to G. G. Stokes, 12 April 1890, GGS-CUL P417.

153. Pole, "On the Present State of Knowledge," 444.

154. William Pole to G. G. Stokes, 16 March 1860, GGS-CUL P411.

155. Pole, "Daltonism," especially 825–30.

156. Ibid., 827, 831–33.

157. J. Herschel, "Light," 435–37; *Helmholtz's Treatise on Physiological Optics*, 2:141–44.

158. Ladd-Franklin, *Colour and Colour Theories*, 191–92.

159. Ash, "Historicizing Mind Science."

CHAPTER 4. MENTAL GOVERNANCE AND HEMIOPSY

1. Descriptions of these attacks can be found in Schiller, "Migraine Tradition"; Levene, "Sir G. B. Airy"; Wade, *Brewster and Wheatstone on Vision*, 226–34; Sacks, "Scotoma."

2. Wise with Smith, "Work and Waste." In subsequent footnotes in this chapter, I will indicate which paper is cited with I, II, or III. Quote above: I:263. Wise and Smith *do* briefly discuss mental-moral philosophy (I:270–71, 286–87; III:224–46), but the analysis can be much extended.

3. Wise with Smith, "Work and Waste," I:268.

4. Broussais, *Traité de physiologie*, 267, 282, 286, 300; cited in Canguilhem, *Normal and the Pathological*, 54-55; Staum, *Cabanis*, 174-206; Locke, *Essay concerning Human Understanding*, 229; Hartley, *Observations on Man;* Hume, *Essays*, 166; *Autobiography of Benjamin Franklin*, 102. Hartley and Hume are cited in Smith, *Norton History of the Human Sciences*, 246-54. Wise and Smith also note associationism's use of the balance model, "Work and Waste," I:271.

5. T. M. Porter, *Rise of Statistical Thinking;* Krüger, *Probabilistic Revolution;* Daston, *Classical Probability in the Enlightenment;* Hacking, *Taming of Chance;* Wise, *Values of Precision.*

6. On "aperspectival objectivity" as a product of the later nineteenth century see Daston, "Objectivity and the Escape from Perspective."

7. Schaffer, "Machine Philosophy,"175; Ashworth, "Memory, Foresight, and Production"; Berg, *Machinery Question.*

8. H. Airy, "On a Distinct Form."

9. J. Herschel, "On Sensorial Vision," 401.

10. Quoted in Brewster, *Memoirs of the Life, Writings, and Discoveries of Sir Isaac Newton*, 1:236-39. This letter is also reprinted in Newton, *Correspondence of Isaac Newton*, 153-54. Brewster claims the first published notice of Newton's "phantasm" appeared in King, *Life of John Locke*, 1:404-8. For Newton's recipe for a "Medicine to cleare the eye-sight," see Iliffe, "Isaac Newton," 152.

11. Buffon, "Dissertation sur les couleurs accidentelles." Other early accounts of accidental colors include: Boyle, *Experiments and Considerations Touching Colours*, 11-20; de La Hire, "Dissertation sur les differens accidens de la vue"; Jurin, "Essay upon Distinct and Indistinct Vision"; Scherffer, *Dissertatione de Coloribus accidentalibus;* Aepinus, "De coloribus accidentalibus"; Fothergill, "Remarks on the Sick Headach"; Haüy, *Traité èlémentaire de physique.*

12. R. W. Darwin, "New Experiments," 314, 329-32. Desmond King-Hele made a convincing case that Erasmus Darwin actually did most of the research for this paper. Filled with ambition for his son, Erasmus seems to have passed the research along to Robert to finish so the latter could earn a reputation and election to the Royal Society. See King-Hele's *Doctor of Revolution*, 176-78; *Letters of Erasmus Darwin*, 67. Erasmus reprinted his son's "findings" in his *Zoonomia*, 2:328-74 (section XL).

13. Castle, *Female Thermometer*, chaps. 9 and 10; Sarbin and Juhasz, "Historical Background."

14. Wise with Smith, "Work and Waste," III:221. On the popularity of the steam engine metaphor at midcentury, see Wise and Smith, "Measurement, Work, and Industry," especially 152-60; Wise, "Mediating Machines," especially 79-92.

15. For instance, see the thermodynamic model of animal chemistry as described in Kremer, *Thermodynamics of Life;* or the modeling of the body on a steam engine by E.-J. Marey in the late nineteenth century, described in Douard, "E.-J. Marey's Visual Rhetoric." Also see Russett, *Sexual Science*, chap. 4.

16. Inkster, *Science and Technology in History*, 12, 67.

17. Ure, *Philosophy of Manufactures*, 55. The first edition appeared in 1835.

18. Ibid., 311. Also see pages 7-8, 20-23, 214-15.

19. Southey, *Sir Thomas More*, 1:171; quoted in Ure, *Philosophy of Manufactures*, 277–78.

20. Berg, *Machinery Question*, 9–19 and *passim;* Sussman, *Victorians and the Machine*.

21. *Memoirs and Letters of Sara Coleridge*, 1:164; quoted in R. Porter and D. Porter, *Sickness and Health*, 51. Also see Rabinbach, *Human Motor*.

22. [Brewster], "Accidental Colours."

23. Babbage, *Economy of Machinery and Manufactures*, 27.

24. Babbage, *Word to the Wise*, quoted in Hyman, *Charles Babbage*, 86–87.

25. Wise, "Precision," 227; T. M. Porter, "Precision and Trust."

26. For example, see Carnot, *Essai sur les machines en general*.

27. Burke, "Bursting Boilers."

28. Babbage, *Economy of Machinery and Manufactures*, 27. George Airy also took an interest in this problem in his "On the Numerical Expression."

29. "Achard's Patent Automatical Regulator," CW-KCL Box 6.

30. Quoted in Dickinson, *Matthew Boulton*, 116–17.

31. [Playfair], "Relation of the Food of Man to His Muscular Power," 322.

32. W. B. Carpenter, *Principles of Mental Physiology*, vii–xxiv, 1–28, and *passim;* Tait, *Lectures on Some Recent Advances*, 24–26.

33. Lewes, *Physical Basis of the Mind*, 67–69, 363–75, and *passim*.

34. W. B. Carpenter, *Principles of Mental Physiology*, 389.

35. Helmholtz, "Recent Progress of the Theory of Vision," 227, 232–33; Helmholtz, "Über die Erhaltung der Kraft," 1:12–75; Brain and Wise, "Muscles and Engines."

36. Morus, "Correlation and Control," 597, 610.

37. Huxley, "Liberal Education," 86. Huxley delivered this lecture in 1868.

38. Haley, *Healthy Body;* Flynn, "Running Out of Matter."

39. Smee, *Vision in Health and Disease*, 30.

40. John Phillips to William Wallace Currie, 7 November 1837, Fitzwilliam Museum, Perceval bequest, J.123; Farrar, Farrar, and Scott, "Henrys of Manchester"; both cases discussed in Morrell and Thackray, *Gentlemen of Science*, 128–36, 402.

41. [Balfour], "Professor George Wilson," 234.

42. Tyndall, *Faraday as a Discoverer*, 75–78; Faraday to John Herschel, 14 February 1839, in *Selected Correspondence of Michael Faraday*, 1:335. Dalton feared Faraday might have suffered lead poisoning, as he thought he had himself four decades earlier. See Dalton to Faraday, 3 September 1840, Royal Institution, printed in Thackray, *John Dalton*, 171–72.

43. John Couch Adams to George Airy, 8 May 1851, RGO-CUL 6/373.180.

44. Clerke, *Modern Cosmogonies*, 160; Huggins, *Agnes Mary Clerke*, 18–20.

45. Edward Smedley to Charles Babbage, 28 October 1827, CB-BL add. ms. 37184.96.

46. William Thomson to George Gabriel Stokes, 16 January and 14 February 1861, GGS-CUL K122 and K124. On the standards committee, see Hunt, "Ohm is Where the Art Is."

47. Galton, *Inquiries into Human Faculty*, 17–19; Galton, *Memories of My Life*, 78–79. Also see Pearson, *Life, Letters and Labours of Francis Galton*, 1:166–73.

48. Babbage, *Economy of Machinery and Manufactures*, 30.

49. The *Oxford English Dictionary* (2nd ed.) does not include an entry for "hemiopsy."

50. William Hyde Wollaston, "Punctum caecum . . . ," n.d., detached leaf in Notebook 5 (Tours), WHW-CUL, Box 1.

51. Wollaston, "On the Semi-decussation of the Optic Nerves."

52. Newton, *Opticks*, 346. This page corresponds to the third book, query 14. Also see David Brewster's discussion of this theory and reproduction of a longer, unpublished description by Newton, in Brewster, *Memoirs of the Life, Writings, and Discoveries of Sir Isaac Newton*, 1:226-36, 432-36. John Taylor also proposed semi-decussation in his *Account of the Mechanism of the Eye*. See Levene, "Sir G. B. Airy," 16.

53. Wollaston, "On an Improvement."

54. Hill, "Life in the Laboratory," 239.

55. "Appearances Observed on Inspecting the Body of the Late William Hyde Wollaston M.D., F.R.S., December 24, 1828," Envelope B, WHW-CUL, Box 2; J. H. Jackson, "Commentary on a Case of Hemiplegia"; Brodie, *Psychological Inquiries*, 125-27.

56. William Hyde Wollaston, "Fordyce's Lectures on Chronic Diseases" and "Quantity of Air Expired . . . ," both in Notebook C; Patient log, Notebook 7, all in WHW-CUL, Box 1.

57. On the transition from kinematics to dynamics, see Smith and Wise, *Energy and Empire*, 192-202, 360-80.

58. Henry Warburton(?), biographical notes, Envelope B; and "Diary" notebook, both in WHW-CUL, Box 2.

59. Arago, "Thomas Young," in *Biographies*, 280. Arago read this éloge to the Académie des Sciences on 26 November 1832.

60. W. Henry, *Elements of Experimental Chemistry*, viii-x. On the 1820 Royal Society election and 1827 Wollaston committee, see Miller, "Between Hostile Camps"; Gleason, *Royal Society*, 80-123.

61. Babbage, *Reflections*, 203-12.

62. William Henry to Charles Babbage, 19 May 1830, CB-BL add. ms. 37185.179-80.

63. Yeo, "Genius, Method, and Morality."

64. Faraday, "Observations on the Education of the Judgment," 206-7, 215-19. Faraday here cites Samuel Butler's poem about human failings, *Hudibras*, part 1, canto 1, line 297.

65. For example, see Twining, "On Single Vision"; Alison, "On Single and Correct Vision," especially 480-84. In a footnote (p. 481), Alison mentions his own hemionopic attacks. Two decades later, this highly prominent Edinburgh physician and former student of Dugald Stewart began having regular seizures (see Stephen, *Dictionary of National Biography*).

66. [Kater], "An Account of Some Curious Facts Respecting Vision"; Heineken, "On a Singular Irregularity of Vision."

67. Brewster, "On Hemiopsy, or Half-Vision."

68. Brewster, *Letters on Natural Magic*, 48.

69. David Brewster to John Herschel, 28 January 1822, JH-RSL 4.255. On the strenuous nature of Brewster's career and of pursuing a scientific life in Edinburgh generally, see Shapin, "Brewster and the Edinburgh Career in Science," and Brock, "Brewster as a Scientific Journalist."

70. J. H. Gladstone, "Sir David Brewster."

71. [Brewster], "Dr. George Wilson," 332.

72. [Brewster], "Accidental Colours," 85.

73. Fechner, *Nanna;* Fechner, *Zend-Avesta;* Fechner entry in Gillispie, *Dictionary of Scientific Biography.*

74. Brewster, "On the Effect of Compression and Dilatation upon the Retina," 92. Brewster's confidence in his authority with respect to vision despite his own impaired sight is evident in his prolific publication on the subject.

75. Stewart, *Lectures on Moral Philosophy,* 198–99.

76. Brewster, "On the Optical Phenomena."

77. Airy, *Autobiography,* 1–2. George wrote most of this text and in that sense it is an autobiography, but Wilfrid organized the text and added comments of his own.

78. Ibid., 5, 8–9, 12–13, 73–74, 346–50 (on Airy's orderly life and balanced mind); 25–26 (on exercise). On mathematics as physical exertion, see Warwick, "Exercising the Student Body."

79. For example, see W. C. Henry, *Memoirs of the Life,* 42–43, 208–10; Struve, *Wilhelm Struve,* 63; *Letters of John Stuart Mill,* 1:xxvi–xxvii.

80. Airy, *Autobiography,* 127–29; R. W. Smith, "National Observatory," 7–8.

81. George Airy, Astronomer Royal's journal, 4 vols., RGO-CUL 6/24-27. See, for example, the entry for 14 January 1837, 24. On the Board's wish to have vision tests made and published, see John Herschel to Francis Baily, 10 February 1831, JH-RSL 25.1.19.

82. The Breen–Airy exchange took place between 1–4 January 1847, and is in RGO-CUL 6/35.2–5. I thank Anne Secord for the last point in this paragraph.

83. George Airy to Jean-Baptiste Biot, 16 December 1850, RGO-CUL 6/372.68–69. Airy hired Charles May, the same man who had assisted Herschel in his color blindness work, to build the transit circle. On the Greenwich observatory as a factory, see Schaffer, "Astronomers Mark Time"; R. W. Smith, "National Observatory," 13–17; Chapman, "Sir George Airy"; Newcomb, *Reminiscences of an Astronomer,* 287–88. On the skills required in factory settings, see Wise and Smith, "Measurement, Work, and Industry," 170; Thomson, "Bangor Laboratories," 2:485; Penn, *Class, Power, and Technology.*

84. Otto Struve to George Airy, 28 July 1859, RGO-CUL 6/378.495–96.

85. James South to the Lord Commissioners of the Admiralty, 12 March 1847, RGO-CUL 6/2.24; George Airy to the Admiralty, 26 March 1847, RGO-CUL 6/2.42; Meadows, *Recent History,* 3.

86. George Airy to J. McCarthy, 23 July 1879, RGO-CUL 6/264.218–19.

87. On Airy's discovery, correction, and monitoring of his astigmatism, see his "On a Peculiar Defect in the Eye"; G. Airy, "On a Change in the State of an Eye"; G. Airy, "On the Continued Change in an Eye"; G. Airy, "Further Observations on the State of an Eye"; Levene, "Sir George Biddell Airy."

88. G. Airy, *Autobiography,* 35, 65, 362–63.

89. G. Airy, "On Hemiopsy." Levene contextualizes this paper well in his "Sir G. B. Airy."

90. G. Airy, "On Hemiopsy," 19–20.

91. G.-H. Dufour to George Airy, 24 April 1868, printed in H. Airy, "On a Distinct Form," 250–51. Hubert Airy mentions Dufour's face washing in a letter to John Herschel, 8 May 1868, JH–RSL 19.308. Hubert did not find the technique helpful in preventing headache.

92. Brewster, "On a New Property of the Retina," 328, 329.

93. G. Airy, "On Hemiopsy," 21.

94. Chapman, "Private Research and Public Duty," 124–30; G. Airy, *Autobiography*, 183.

95. George Airy to Christopher Hansteen, 29 March 1865, RGO–CUL 6/381.416–17. For a discussion of Darwin's health as reflected in his *cartes de visite*, see J. Browne, "I Could Have Retched All Night."

96. Caroline Herschel to John Herschel, 14 July 1823, CH–BL Egerton ms. 3761; David Brewster to John Herschel, 2 December 1825, JH–RSL 4.259.

97. J. Herschel, *Preliminary Discourse*, 3.

98. John Herschel to Charles Babbage, 19 March 1820, JH–RSL 2.130.

99. Agassi, "Sir John Herschel's Philosophy of Success."

100. [J. Herschel], "Whewell on the Inductive Sciences," 182.

101. John Herschel to George Airy, 6 September 1852, RGO–CUL 6/466.502.

102. George Airy to John Herschel, 7 September 1852, RGO–CUL 6/466.503–4. Hemiplegia is "paralysis on one side of the body, usually caused by a lesion in the opposite side of the brain," according to the *Oxford English Dictionary* (2nd ed.). Hemiplegia was one of many morbid conditions presumed to cause subjective visions. See, e.g., J. H. Jackson, "Commentary on a Case of Hemiplegia."

103. John Herschel to George Airy, 9 November 1862, RGO–CUL 6/380.265–66. In this letter, Herschel describes cerebral spectra brought on by working with fungi spores.

104. John Herschel diary, 11 June 1846, JH–RSL MS.584. Also see entries for 11 May 1853, 20 February 1854, 18 August 1855, 7 February 1858, 12 February 1865, 29 July 1866, 24 February and 5 May 1868; 22 June, 7 August, and 11 October 1869; 16 April, 17 July, and 5 September 1870; "Optical Mss.—Unidentified Material," 4 January 1852, JH–HRC W0354; "Sensorial Vision," 406–7.

105. J. Herschel, "Light"; J. Herschel, "On the Action of Crystallized Bodies."

106. John Herschel, "Miscellanea," 12 May 1820, JH–HRC M-W0070; Herschel and South, "Observations of the Apparent Distances," 15.

107. John Herschel to unknown recipient (possibly his brother-in-law James Calder Stewart), n.d. (but after 1827), labeled "The Eye," JH–HRC M-W0067. Herschel noted reading of a similar experiment by the surgeon T. Smith, who read a paper to the Royal Society of Edinburgh on 3 April 1826. For this note, see John Herschel, "Optical Mss.—Unidentified Material," n.d., JH–HRC W0354.

108. Buttmann, *Shadow of the Telescope*, 181–82. Laudanum was the tincture of opium and used widely without regulation until 1868. Herschel acknowledged that the drug might have exacerbated his visions, but remained convinced that the true cause lay elsewhere. He seems justified in this belief, for his visions began before he started taking laudanum.

109. Herschel, "Optical Mss.—Unidentified Material," JH–HRC W0354.

110. John Herschel to Mary Somerville(?), n.d., JH–HRC L0760. On Herschel's ill health, also see George Airy to Herschel, 3 February 1839, JH–RSL 1.76; Herschel to Airy, 16 April 1856(?), JH–RSL 1.331; Francis Baily to Herschel, 9 January 1841, JH–RSL 3.197; Herschel to Baily, 1 March 1837 and 26 January and 11 October 1841, JH–RSL 3.138, 198, 218; Isabella Herschel to Edward Sabine, 6 April 1855, ES–RSL MS.258.639; Margaret Herschel to J. W. Lubbock, 20 October 1855/6(?), JWL–RSL H.364; John Herschel diary, JH–RSL MS.584–86; Buttmann, *Shadow of the Telescope*, 154–55.

111. Isabella Herschel to George Airy, 23 March 1855, RGO–CUL 6/336.35.

112. "Sir John Frederick William Herschel."

113. Maria Edgeworth to M. Pakenham Edgeworth, 26 November 1843, in *Maria Edgeworth*, 594–95. Herschel made the photography/hemiopsy analogy explicitly in "Sensorial Vision," 402–3. On an earlier occasion, Herschel had explained Newton's theory of fits to his guest by building an elaborate analogy of two hypochondriacs trying to walk through a door. See Maria Edgeworth to Harriet Butler, 29 March 1831, in *Maria Edgeworth*, 501–2. On Edgeworth's views about nervous men, see Logan, *Nerves and Narratives*, chap. 5.

114. J. Herschel, "Sensorial Vision," 407, 411–12.

115. Ibid., 402 (emphasis mine). Oddly, even when Airy received Herschel's *Familiar Lectures*, which contained a reprint of the Leeds talk, the Astronomer Royal did not comment on the similarities between their visions (Airy to Herschel, 29 January 1867, JH–RSL 1.304).

116. John Herschel to William B. Carpenter, 23 November 1858, JH–RSL 5.195; Sophia de Morgan to John Herschel, n.d., JH–RSL 6.434.

117. Hubert Airy to John Herschel, 30 April 1868, JH–RSL 19.307. On the 1854 attack, also see H. Airy, "On a Distinct Form," 255.

118. Hubert Airy here employed the principle behind the camera lucida. This instrument used prisms to "project" an image (of the scene in front of the observer) onto a piece of paper. The observer then traced the image. See Schaaf, *Tracings of Light*.

119. Airy's half of this correspondence, which spanned from 29 February to 18 November 1868, can be found in JH–RSL 19.306–10. Two of Herschel's responses are excerpted in H. Airy, "On a Distinct Form," 252–53. Airy seems to have misquoted the date of Herschel's second letter as 17 November 1869; it should read 1868.

120. John Herschel to Hubert Airy, 4 May 1868, excerpted in H. Airy, "On a Distinct Form," 252.

121. For example, London Hospital physician J. Hughlings Jackson had just published an analysis of such a case. See his "Commentary on a Case of Hemiplegia."

122. Fothergill, "Remarks on the Sick Headach"; Tyrrell, *Practical Work on the Diseases of the Eye; Collections from the Unpublished Medical Writings of the Late Caleb Hillier Parry*, 557–58; H. Airy, "On a Distinct Form," 254–55.

123. H. Airy, "On a Distinct Form," 250. George Airy communicated this version of his son's thesis to the Royal Society on 17 February 1870. Hubert was not satisfied with the term *hemiopsy*, since it implied a return to the discredited semi-decussation theory. See his letter to John Herschel, 8 May 1868, JH–RSL 19.308.

124. H. Airy, "On a Distinct Form," 259–64.
125. George Airy, Astronomer Royal's journal, RGO–CUL 6/26, entry for 17 February 1870.
126. Charles Wheatstone, 10 January 1870, RR–RSL 7.4.
127. Ibid.; Charles Wheatstone, "Mental Spectra," n.d. and 18 November 1850, CW–KCL Section 4, Box 2, File 12; Hubert Airy to G. G. Stokes, 19 January 1870, GGS–CUL A572.
128. H. Airy, "On a Distinct Form," 247.
129. W. B. Carpenter, *Principles of Mental Physiology*, 546–47; Hawksworth, *Workshop of the Mind*, 93–94.
130. [Brewster], "Phrenology," 70.
131. J. Herschel, *Preliminary Discourse*, 133–34; Cawood, "Magnetic Crusade"; Yeo, "Idol of the Marketplace."
132. J. Herschel, *Preliminary Discourse*, 77. Also see J. Herschel, "1845 Presidential Address," 651.
133. Whewell, *History of the Inductive Sciences*, 1:20–21; Tyndall, *Mountaineering in 1861*, 6; Tyndall, *Heat Considered as a Mode of Motion*, 137.
134. George Airy to John Herschel, 10 October 1865, JH–RSL 1.283.
135. Galison, *How Experiments End*, 244. Also see Buchwald, *Scientific Practice*; Biagioli, "Tacit Knowledge."
136. George Airy to John Herschel, 16 July 1831, JH–RSL 1.49.
137. Barker-Benfield, *Culture of Sensibility*, especially 215–86; Russett, *Sexual Science*, 104–29.
138. [Brewster], "François Arago," 493–94.
139. Oreskes, "Objectivity or Heroism?" In the same volume, see Hevly, "Heroic Science of Glacier Motion."

CHAPTER 5. RATIONAL FAITH AND HALLUCINATION

1. *Life of Mary Anne Schimmelpenninck*, 1:86–89; Ingram, *Haunted Homes*, 20–21.
2. Carlyle, "Signs of the Times," 34; Lang, *Cock Lane and Common-Sense*, 28, 30; Pearsall, *Table-Rappers*, 139; Dickerson, *Victorian Ghosts in the Noontide*, chap. 1; Castle, *Female Thermometer*.
3. Clery, *Rise of Supernatural Fiction*; Voller, *Supernatural Sublime*; Carter, *Spectre or Delusion*.
4. Dickerson, *Victorian Ghosts in the Noontide*, 14.
5. Daston and Park, *Wonders and the Order of Nature*; Daston, "Marvelous Facts." Also see Alderson, *Essay on Apparitions*; Clark, "Scientific Status of Demonology"; Brooke, *Science and Religion*, chap. 4.
6. Dear, *Discipline and Experience*, chap. 1.
7. Berrios, *History of Mental Symptoms*, 60 (emphasis in the original). On the medicalization and symptomatization of hallucinations, also see Sarbin and Juhasz, "Historical Background."

8. Hole, *Pulpits, Politics, and Public Order;* Hempton, *Religion and Political Culture.*

9. On how Protestant versus Catholic notions of miracles have shaped the epistemology of experiments, see Dear, "Miracles, Experiments, and the Ordinary Course of Nature."

10. For example, see Whiston, *Account of the Exact Time;* Daston, "Marvelous Facts," 114–15; Burns, *Great Debate on Miracles.*

11. Dendy, *Philosophy of Mystery,* 334.

12. "Ferriar on *Apparitions,*" 312.

13. Hempton, *Religion and Political Culture,* 12. On the pathologizing of enthusiasm, see R. Porter, "Rage of Party"; Hawes, *Mania and Literary Style.*

14. Cannon, *Science in Culture,* 61.

15. Hole, *Pulpits, Politics, and Public Order,* 57–59, 67–72, 97–98, and *passim;* Brooke, *Science and Religion,* chaps. 5 and 6. Hume argued that religious factions would "seduce silly women and ignorant mechanics into a belief of the most monstrous principles." See his *History of England,* 4:122; Hume, "On Miracles," in *Essays.*

16. Morrell and Thackray, *Gentlemen of Science,* 224–45; Inkster and Morrell, *Metropolis and Province;* Cannon, *Science in Culture;* Inkster, "London Science."

17. J. Henry, "Occult Qualities," 338.

18. Glanvill, *Saducismus Triumphatus;* Clery, *Rise of Supernatural Fiction,* 18–24.

19. Paley, *View of Evidences of Christianity;* Paley, *Natural Theology;* Fyfe, "Reception of William Paley's *Natural Theology*"; Gillespie, "Divine Design"; Hole, *Pulpits, Politics, and Public Order,* 73–82.

20. Gascoigne, "From Bentley to the Victorians"; Hahn, "Laplace and the Mechanistic Universe"; Brooke, "Scientific Thought," 44–48.

21. Knox, "Dephlogisticating the Bible"; Ashworth, "Memory, Foresight, and Production," chap. 1.

22. Whewell, *Astronomy and General Physics,* 350; quoted in Brooke, "Scientific Thought," 46.

23. Reid, *Essays on the Active Powers of the Human Mind,* 46; Heimann, "Voluntarism and Immanence," 278–79.

24. Brooke, *Science and Religion,* 134–40; Smith and Wise, *Energy and Empire,* 85–87.

25. J. Davy, *Memoirs of the Life of Sir Humphry Davy,* 1:15–16.

26. Schaffer, "States of Mind"; Schofield, "Joseph Priestley," 248; Brooke, *Science and Religion,* 177.

27. Olson, *Science Deified and Science Defied,* 2:119–20, 136.

28. La Mettrie, *L'Homme Machine,* 62 (emphasis in original).

29. W. B. Carpenter, *Principles of Mental Physiology,* 691–92.

30. J. Herschel, *Preliminary Discourse,* 144; C. Smith, "Mechanical Philosophy"; Schofield, *Mechanism and Materialism,* chaps. 10, 11; Thackray, *Atoms and Powers.*

31. Peterfreund, "Scientific Models," 59–73; Cantor, "Henry Brougham"; Morse, "Natural Philosophy."

32. Cantor, "Theological Significance of Ethers."

33. [Brewster], "Bridgewater Bequest," 429; Olson, *Scottish Philosophy and British Physics,* chap. 7; Morse, "Natural Philosophy."

34. John Herschel diary, 10 January 1845 and 12 May 1848, JH-RSL MS.584–85.

35. Michael Faraday to the editor of the *Times*, 28 June 1853, in *Selected Correspondence*, 2:690-92; Cantor, *Michael Faraday*; Gooding, "Metaphysics versus Measurement."

36. Campbell and Garnett, *Life of James Clerk Maxwell*, 156.

37. James Clerk Maxwell, "Idiotic Imps," summer term, 1853, in Campbell and Garnett, *Life of James Clerk Maxwell*, 227-30.

38. Walker, "Description of Mr. Ez. Walker's New Optical Machine." Also see "Ferriar on *Apparitions*," 306.

39. Castle, "Phantasmagoria," 58. A revised version of this paper appears in *Female Thermometer*, chap. 10. On nineteenth-century popular entertainments, see Hankins and Silverman, *Instruments and the Imagination*; Altick, *Shows of London*.

40. Dendy, *Philosophy of Mystery*, 56.

41. Locke, *Essay concerning Human Understanding*, 562-63; also see 372, 394-95.

42. Holland, *Chapters on Mental Physiology*, 114-15; Brodie, *Psychological Inquiries*, 81-91.

43. Daston, "Fear and Loathing"; Tuveson, *Imagination as a Means of Grace*; McNeil, *Under the Banner of Science*, chap. 2.

44. [T. Brown], "Stewart's *Life of Dr. Reid*," 282; Blakey, *History of the Philosophy of Mind*, 3:189-208.

45. Stewart, *Lectures on Moral Philosophy*; John Lee, Notes from Stewart's moral philosophy course, 1796-1799, DS-EUL Dc.8.143. Also see Arnold, *Observations*, 1:76.

46. John Herschel, "Cause and Effect," 27 February 1842, JH-HRC M-W0062; "Philosophy Notes," n.d., JH-HRC M-W0064. Also see his scathing attack on Kant's, Fichte's, and Schelling's metaphysics in his letters to his wife, Margaret Herschel, n.d. 1838, JH-HRC L0538.

47. Lavater, *Of Ghostes and Spirites*, 1; discussed in Sarbin and Juhasz, "Historical Background," 345.

48. T. Browne, *Pseudodoxica Epidemica*, 2:221; Defoe, *Essay on the History and Reality of Apparitions*; both discussed in Castle, "Phantasmagoria."

49. E. Carpenter, *Nocturnal Alarm*; Watkins and Shoberl, *Biographical Dictionary*, 55.

50. Crosland, *Apparitions*, 60.

51. Crichton, *Inquiry into the Nature and Origin of Mental Derangement*.

52. Arnold, *Observations*, 1:55-58.

53. Esquirol, *Mental Maladies*, 93-110. The original work, *Des Maladies Mentales Considérées sous les Rapports Médical Hygienique et Médico-Legal*, appeared in 1838.

54. Ferriar, *Essay towards a Theory of Apparitions*, 31-32, 109-10, 117, 123-24, 137-39. Also see Ferriar, "Of Popular Illusions," 31-32.

55. Nicolai, "Memoir on the Appearance of Spectres."

56. Ibid.

57. Hibbert, *Sketches of the Philosophy of Apparitions*, 3-8. For the organic and often sane nature of hallucinations, see 19-20, 342, and *passim*.

58. Ibid., 21-25, 55-79, 86-88.

59. Ibid., 422.

60. Abercrombie, *Inquiries concerning the Intellectual Powers*, 60, 77-79, 137, 164, and *passim*. On Abercrombie, see Chambers, *Biographical Dictionary of Eminent Scotsmen*, 1:1-3.

61. Newnham, *Essay on Superstition*, 46, 49–81, and *passim*.

62. Ibid., 3–4, 12, 21–24, 112–13.

63. Ibid., 19–20, 342.

64. Ibid., 4–7.

65. Ibid., 8.

66. Ibid., 119–21.

67. Dendy, *Philosophy of Mystery*, 33–36, 55.

68. Ibid., 102, 172–77.

69. Ibid., 168–69. Newnham also expounded this view in his *Essay on Superstition*, 35–37.

70. Dendy, *Philosophy of Mystery*, 64–66.

71. Lockhart, *Memoirs of Sir Walter Scott*, 9:334–38; Scott, *Letters on Demonology*, 10; Walter Scott to Louisa Stuart, 31 October 1830, in *Letters of Sir Walter Scott*, 11:400–404.

72. Walter Scott to J. G. Lockhart, 21 March 1830, in *Letters of Sir Walter Scott*, 11:309–10.

73. Scott, *Letters on Demonology*, 320; also see 154–60, 285–87. For Scott's views on superstition, see Cottom, *Civilized Imagination*, chap. 8.

74. Parsons, *Witchcraft and Demonology in Scott's Fiction*, 178.

75. Scott, *Letters on Demonology*, 317.

76. Ibid., 45–46.

77. Ibid., 36–37. Scott credited this story to Thiebault, *Original Anecdotes of Frederic the Second*. Though Thiebault discussed Maupertuis and Gleditsch several times, he did not tell this particular story.

78. Scott, *Letters on Demonology*, 306–8.

79. Ibid., 29–34. I have not been able to determine who this scientific gentleman was.

80. Cottom, *Civilized Imagination*, 149.

81. Brewster, *Letters on Natural Magic*, 1–2. On Brewster and Scott's frequent meetings during the late 1820s, see *Journal of Sir Walter Scott*, 152, 158, 273, 290, 307, 370, 444–45, 455, 459.

82. Brewster, *Letters on Natural Magic*, 5, 10.

83. Ibid., 29. Also see 11–20, 48–49.

84. Ibid., 46–47. An earlier discussion of Mrs. A. appeared as "Account of a Remarkable Case of Spectral Illusion."

85. Brewster, *Letters on Natural Magic*, 3. Also see [Brewster], "Works on Mental Philosophy."

86. Somerville, *Personal Recollections*, 66.

87. Gordon, *Home Life of Sir David Brewster*, 16–18.

88. Brewster, *Letters on Natural Magic*, 48–51. On the Hibbert incident, see Gordon, *Home Life of Sir David Brewster*, 133–34.

89. Brewster, *Letters on Natural Magic*, 127–28, 147–48. Brewster probably learned of this explanation from Jordan, "Observations on a Singular Phenomenon." The Saxon idolatry reference appears in Dendy, *Philosophy of Mystery*, 68. I have not been able to determine who "Haue" was. Perhaps Jordan and Brewster meant Viennese paleontologist Joseph von Hauer (1778–1863).

90. Brewster, *Letters on Natural Magic*, 350; Baxter, "Brewster, Evangelicalism and the Disruption of the Church of Scotland," 46-47.

91. "Passages from the Diary of a Late Physician, ch. IV," 786.

92. Brierre de Boismont, *Hallucinations*, vi-vii.

93. Ibid., 46-74.

94. Ibid., 21. Also see 35, 102, 117, 285-348, 357, 360, 449-53, 496, and *passim*.

95. Ibid., 125, 401.

96. Paley, *Natural Theology*, 18-38.

97. James Jago, "Ocular Spectres and Structures as Mutual Exponents," PA-RSL 36.24.

98. Chalmers, *On the Power, Wisdom, and Goodness of God*, 2:159; J. Herschel, "Man, the Interpreter of Nature," 737; Yeo, "William Whewell," 497-98.

99. Hibbert, *Sketches of the Philosophy of Apparitions*, 72-75. Also see Siraisi, *Clock and the Mirror*.

100. Hibbert, *Sketches of the Philosophy of Apparitions*, 355-56.

101. Van Helmont, *Ortus medicinae*; Brierre de Boismont, *Hallucinations*, 205.

102. Taylor, *Apparitions*, 222-23.

103. Dendy, *Philosophy of Mystery*, 81.

104. Ibid., 21-24.

105. Winslow, *Anatomy of Suicide*, 123, 126.

106. *Goethe's sammtliche Werke*, 22:83.

107. Cabanis, *On the Relations between the Physical and Moral Aspects of Man*.

108. Bonnet, *Essai analytique sur les Facultés de l'áme*, 14:176-77; Levêsque de Pouilly, *Éloge de Charles Bonnet*, 120-21.

109. Ebbecke, *Johannes Müller*, 88-89.

110. Dendy, *Philosophy of Mystery*, 81. On Conolly, see Scull, "Victorian Alienist," 1:103-50; Showalter, *Female Malady*, chap. 1.

111. On Sandras, see "Reports of the Sessions," 541-44; discussed in Berrios and Dening, "Pseudohallucinations," 755-56; Sandras, *Traite pratique des maladies nerveuses*. On Gruthuisen, see Brierre de Boismont, *Hallucinations*, 426.

112. *Compte Rendu des Séances de l'Académie des Sciences* 22 (1846): 306.

113. "Séances des 9 et 16 février," 1; Rogers, *Philosophy of Mysterious Agents*, 56; Crowe, *Night Side of Nature*, 380.

114. Brierre de Boismont, *Hallucinations*, 178-79.

115. Crichton, *Inquiry into the Nature and Origin of Mental Derangement*, 2:35.

116. Ibid., 1:132; Hibbert, *Sketches of the Philosophy of Apparitions*, 305-6.

117. "On the Phantasms Produced by Disordered Sensation"; quoted in Hibbert, *Sketches of the Philosophy of Apparitions*, 49-54, 208-10, 323-24.

118. Bostock, *Elementary System of Physiology*, 3:204; discussed in Conolly, *Inquiry concerning the Indications of Insanity*, 112, and Brierre de Boismont, *Hallucinations*, 56-57.

119. Davy, *Researches, Chemical and Philosophical*, 269-330; Hoover, "Coleridge, Humphry Davy, and Some Early Experiments"; Tuttle, "Humphry Davy," chap. 3.

120. De Quincey, *Confessions of an English Opium Eater*; Maehle, "Pharmacological Experimentation"; Hayter, *Opium and the Romantic Imagination*; Logan, *Nerves and Narratives*, chap. 4.

121. John Herschel, "Optical Mss.—Unidentified Material," n.d., JH–HRC W0354.

122. J. Herschel, "Sensorial Vision," 404–5.

123. John Herschel diary, 9 November 1847, 2 November 1853, 26 October 1864, and 12 February 1865, JH–RSL MS.584-86.

124. John Herschel to Charles Babbage, 4 August 1814, JH–HRC L0047. I have not been able to locate Babbage's reply.

125. Babbage, *Passages from the Life of a Philosopher*, 7–10, 25.

126. Brodie, *Psychological Inquiries*, 140–41; Charles Babbage to Benjamin Brodie, n.d. November 1854, CB–BL add. ms. 37196.8.

127. Pearson, *Life, Letters and Labours of Francis Galton*, 2:244; Galton, *Inquiries into Human Faculty*, 67.

128. For instance, see G. H. Lewes's defensive essay on the sanity of geniuses, "Great Wits, Mad Wits?"

129. Müller, *Physiology of the Senses*, 1417; Brodie, *Psychological Inquiries*, 141–42, 145, 152–56; Brierre de Boismont, *Hallucinations*, 203. On genius, see Yeo, "Genius, Method, and Morality"; Alborn, "Business of Induction."

130. Galton, "Statistics of Mental Imagery"; Galton, "Mental Imagery"; Galton, "Visions of Sane Persons." Galton gathered these various investigations into his *Inquiries into Human Faculty*, which first appeared in 1883.

131. Francis Galton, unpublished letter to *Nature*, quoted in Forrest, *Francis Galton*, 156–57. Forrest says this letter is kept in FG–UCL under the name "Hallucinations and Half-Hallucinations," but I have not been able to locate this document.

132. Galton, *Inquiries into Human Faculty*, 57.

133. Forrest, *Francis Galton*, 149.

134. Galton, *Inquiries into Human Faculty*, 58. For the original forms and responses, see FG–UCL 152/1, 152/2A, B, and C.

135. "99 Women's Questionnaires on Mental Imagery, 1879–80," FG–UCL 152/2A. For his rankings, see Galton, *Inquiries into Human Faculty*, 61–65. Generally, Galton preserved the anonymity of his subjects in print.

136. Galton, *Inquiries into Human Faculty*, 58–61. Also see Galton's slightly different draft of this portion of his *Inquiries into Human Faculty*, in "Visualizing," FG–UCL 152/1, and "Hallucinations," FG–UCL 152/11; Isabella Herschel to Francis Galton, 8 August 1880(?), FG–UCL 152/5A.

137. John Marshall in "96 Men's Questionnaires on Mental Imagery, 1879–80," FG–UCL 152/2B; John Herschel to Francis Galton, 21 February 1880 and 26 February 1880, FG–UCL 152/4.

138. Galton, "Mental Imagery," 65. Also see Galton, "Visions," 519–20.

139. Galton, *Inquiries into Human Faculty*, 112–28; on Booth and Schuster, see FG–UCL 152/2B; Galton, *Memories of My Life*, 273–74. For other reports of hallucinations from Galton's subjects, see letters from Wilma E. Glanville, 13 April 1884, FG–UCL 152/4; Sophia de Morgan, 2 June 1880(?), and Teresa Lewis, 24 June 1881(?), all in FG–UCL 152/5A; Virginia C. Minor, 30 October 1881, and Charlotte O'Brien, n.d., both in FG–UCL 152/5B.

140. Francis Galton, "Hallucinations," FG–UCL 152/11.

141. Francis Galton, "Mental Imagery," 64; Galton, "Statistics of Mental Imagery"; Galton, *Inquiries into Human Faculty*, 127-28.
142. W. Anderson Smith to Francis Galton, 13 February 1880, FG-UCL 152/2B. Also see Athenaeum librarian Henry R. Tedder's questionnaire in the same folder.
143. Galton projected onto a screen different faces of people from one race, and with a bit of imagination was able to "see" a standard racial figure in the composite image. See Galton, "Mental Imagery," 71-72.
144. Galton, *Inquiries into Human Faculty*, 19-23, 68-73, 76, 79.
145. For example, see Tuke, "Hallucinations"; Inglis, "Hallucinations and Illusions of the Sane"; Colman, "Hallucinations in the Sane"; Seashore, "Measurement of Illusions"; Parish, *Hallucinations and Illusions*.
146. Gurney, Myers, and Podmore, *Phantasms of the Living*, 1:457-573.
147. Sidgwick, "Census of Hallucinations"; Parish, *Hallucinations and Illusions*, 82-109.
148. Marillier, "Statistique des hallucinations."
149. For example, see Maudsley, *Natural Causes and Supernatural Seemings*, chaps. 2 and 3.
150. Symondson, *Victorian Crisis of Faith*; Chadwick, *Victorian Church*, 2:112-50; F. M. Turner, *Between Science and Religion*; Barton, "'Influential Set of Chaps.'"
151. Stewart and Tait, *Unseen Universe*; Heimann, "Unseen Universe."
152. Ferriar, "Of Popular Illusions," 83.
153. "De Foe on Apparitions," 202.

CHAPTER 6. CONCLUSION

1. For example, see Helmstadter and Lightman, *Victorian Faith in Crisis*; Mason, *Making of Victorian Sexuality*; Cain, *Empire and Imperialism*.
2. Holzmann and Pehrson, *Early History of Data Networks*, chap. 2.
3. Actually, investigations into optical telegraphy had a history dating to at least the seventeenth century—among others, Robert Hooke had invented a kind of semaphore device—but Chappe was the first to develop a system for extensive use. See Holzmann and Pehrson, *Early History of Data Networks*, chap. 1; Gamble, *Essay on the Different Modes of Communication with Signals*, 16-56.
4. [Editorial Response], 815.
5. In fact, the Admiralty closed these lines only after the advantages of the electric telegraph became clear in 1847. See Holmes, *The Semaphore*, 103-4.
6. Ibid., 32-41; Holzmann and Pehrson, *Early History of Data Networks*, chap. 5; Gamble, *Essay on the Different Modes of Communication with Signals*.
7. Holmes, *The Semaphore*.
8. For example, see Locke, *Essay concerning Human Understanding*, 397-401.
9. Edelcrantz, *Treatise on Telegraphs*; Rider, "Measure of Ideas," in 125-32. Interestingly, Edelcrantz was an avid, but ultimately unsuccessful suitor to Maria Edgeworth, whose father, Richard Lovell Edgeworth, had been working on his own "tellograph." Edelcrantz also won high acclaim in Britain for his invention of a

safety valve for steam engines. See Holzmann and Pehrson, *Early History of Data Networks*, chap. 3.

10. Holmes, *The Semaphore*, 52–53, 139–40, 155; Holzmann and Pehrson, *Early History of Data Networks*, chap. 2, appendices B, C.

11. For example, see Stewart, *Elements of the Philosophy of the Human Mind*; T. Brown, *Lectures on the Philosophy of the Human Mind*; Holland, *Chapters on Mental Physiology*; Carpenter, *Principles of Mental Physiology*.

12. Holzmann and Pehrson, *Early History of Data Networks*, chap. 2.

13. For example, see Edelcrantz, *Treatise on Telegraphs*.

14. Gamble, *Essay on the Different Modes of Communication with Signals*, 78–91, 96–97; Lenoir, "Helmholtz and the Materialities of Communication"; Morus, "Electric Ariel," 375.

15. Clarke and Jacyna, *Nineteenth-Century Origins of Neuroscientific Concepts*, 157–211.

16. *Extracts from the Private Letters of the Late Sir William Fothergill Cooke*, 9.

17. Spencer, *Principles of Psychology*, 1:386–87. I thank Paul White for this reference.

18. Prescott, *History, Theory, and Practice of the Electric Telegraph*, 242; quoted in Morus, "'Nervous System of Britain,'" 471. For many other examples of the analogy, see the remainder of Morus's article; also, Fahie, *History of Electric Telegraphy*, 239, 303–4; Standage, *Victorian Internet*, 87, 98, 151–52, 160.

19. Standage, *Victorian Internet*, 62–64, 93–95, 122–25.

20. Gooding, "How Do Scientists Reach Agreement?"

21. Aarsleff, *Study of Language in England*.

22. A useful sociological literature on this subject includes Gieryn, "Boundaries of Science"; Easton and Schelling, *Divided Knowledge*; Abbott, *System of Professions*.

23. Cunningham, "How the *Principia* Got its Name"; Dear, "Religion, Science and Natural Philosophy"; Cunningham, "Getting the Game Right"; Cunningham and Williams, "De-centring the Big Picture"; Osler, "Mixing Metaphors."

24. For example, see Brooke, *Science and Religion*, 192–320; Corsi, *Science and Religion*, 227–91.

25. Reed, *From Soul to Mind*, 3.

26. Cannon, *Science in Culture*, 111–36.

27. For an engaging and broad-ranging overview of this change in the human sciences, see R. Smith, *Norton History of the Human Sciences*.

28. Staum, *Cabanis*, 176–77.

29. R. Smith, "Background of Physiological Psychology," 90.

30. Hatfield, "Remaking the Science of Mind"; Robertson, "Bacon-Facing Generation"; Suzuki, "Dualism and the Transformation of Psychiatric Language." Hatfield has argued that psychology was a science as early as the seventeenth century. I am not convinced of the importance of arguing that there was such a thing as science proper in early periods. Regardless, I think Hatfield and I could agree on the different point I wish to make here, which is that psychology began to emerge as a *distinct* discipline around the turn of the nineteenth century.

31. [S. Brown], "Animal Magnetism," 135.

32. Bynum, "Nervous Patient."

Bibliography

MANUSCRIPTS

British Library
  CB-BL Charles Babbage correspondence
  CH-BL Caroline Herschel correspondence
  FN-BL Francis Napier correspondence
  GW-BL George Wilson-Daniel Macmillan correspondence

Cambridge University Library
  RGO-CUL George Airy papers, Royal Greenwich Observatory, Box 6
  JCM-CUL James Clerk Maxwell papers, Ms. Add. 7655
  GGS-CUL George Gabriel Stokes papers, Ms. Add. 7656
  WTK-CUL William Thomson, Lord Kelvin papers, Ms. Add. 7342
  WHW-CUL William Hyde Wollaston papers, Ms. Add. 7736

Edinburgh University Library
  DB-EUL David Brewster correspondence
  JDF-EUL James David Forbes papers
  DRH-EUL David Ramsay Hay correspondence
  DS-EUL Dugald Stewart papers

Harry Ransom Humanities Research Center, University of Texas at Austin
  JH-HRC John Herschel papers

John Rylands University Library of Manchester
  JD-JRL John Dalton papers

King's College London
  JCM-KCL James Clerk Maxwell notebooks
  CW-KCL Charles Wheatstone papers

Royal Society of London
JH-RSL John Herschel papers (microfilm version, ed. Michael Crowe)
JWL-RSL John W. Lubbock correspondence
ES-RSL Edward Sabine correspondence
TY-RSL Thomas Young correspondence
MC-RSL Miscellaneous correspondence
PA-RSL Paper abstracts
RR-RSL Referees' reports

Trinity College, Cambridge University
WW-TC William Whewell papers

University College London
TY-UCL Thomas Young notebooks
FG-UCL Francis Galton papers

Wellcome Library for the History and Understanding of Medicine, London
WLHUM miscellaneous works

Published Sources

Aarsleff, Hans. *The Study of Language in England, 1780–1860*. Princeton: Princeton University Press, 1967.

Abbott, Andrew. *The System of Professions: An Essay on the Division of Expert Labor*. Chicago: University of Chicago Press, 1988.

Abercrombie, John. *Inquiries concerning the Intellectual Powers, and the Investigation of Truth*. Edited by Jacob Abbott. Revised American ed. New York: Robert B. Collins, 1855.

Adas, Michael. *Machines as the Measure of Men: Science, Technology, and Ideologies of Western Dominance*. Ithaca: Cornell University Press, 1989.

Addams, R. "An Account of a Peculiar Optical Phenomenon." *Philosophical Magazine* 5 (1834): 373–74.

Aepinus, Franz Ulrich Theodosius. "De coloribus accidentalibus." *Novum Commentare* 10 (1764): 33–35, 283–88, 292–95.

Agassi, Joseph. "Sir John Herschel's Philosophy of Success." *Historical Studies in the Physical Sciences* 1 (1969): 1–36.

Airy, George. *Autobiography*. Edited by Wilfrid Airy. Cambridge: Cambridge University Press, 1896.

———. "Further Observations on the State of an Eye Affected with a Peculiar Malformation." *Transactions of the Cambridge Philosophical Society* 12 (1879): 392–93.

———. "On a Change in the State of an Eye Affected with Mal-formation." *Transactions of the Cambridge Philosophical Society* 8 (1849): 361–62.

——. "On a Peculiar Defect in the Eye, and a Mode of Correcting It." *Transactions of the Cambridge Philosophical Society* 2 (1827): 267–71.

——. "On Hemiopsy." *Philosophical Magazine* 30 (1865): 19–21.

——. "On the Continued Change in an Eye Affected with a Peculiar Formation." *Proceedings of the Cambridge Philosophical Society* 2 (1864–76): 47–49.

——. "On the Numerical Expression of the Destructive Energy in the Explosions of Steam-Boilers, and on Its Comparison with the Destructive Energy of Gunpowder." *Philosophical Magazine* 26 (1863): 329–36.

——. *Report of the Astronomer Royal to the Board of Visitors of the Royal Observatory, Greenwich.* London, 1853.

Airy, Hubert. "On a Distinct Form of Transient Hemiopsia." *Philosophical Transactions of the Royal Society of London* 160 (1870): 247–64.

Alborn, Timothy L. "The Business of Induction: Industry and Genius in the Language of British Scientific Reform, 1820–1840." *History of Science* 34 (1996): 91–121.

——. "The 'End of Natural Philosophy' Revisited: Varieties of Scientific Discovery." *Nuncius* 3 (1988): 227–50.

——. "Wasted Work: Doctors and Bodies in Early Victorian Life Insurance." Paper presented at the annual meeting of the History of Science Society. Kansas City, October 1998.

Alderson, John. *An Essay on Apparitions; in Which Their Appearance is Accounted for by Causes Wholly Independent of Preternatural Agency.* New ed. London: Longman, Hurst, Rees, Orme, Brown, and Green, 1823.

Alison, William Pulteney. "On Single and Correct Vision, by Means of Double and Inverted Images on the Retinae." *Transactions of the Royal Society of Edinburgh* 13 (1836): 472–93.

Altick, Richard. *The Shows of London.* Cambridge, MA: Belknap Press, 1978.

Altman, Lawrence K. *Who Goes First? The Story of Self-Experimentation in Medicine.* New York: Random House, 1987.

Anderson, Katharine. "Instincts and Instruments." In *The Transformation of Psychology: Influences of 19th-Century Philosophy, Technology, and Natural Science,* edited by Christopher D. Green, Marlene Shore, and Thomas Teo, 153–74. Washington, DC: American Psychological Association, 2001.

——. *Predicting the Weather: Victorians and the Science of Meteorology.* Chicago: University of Chicago Press, 2005.

Appleby, John H. "Human Curiosities and the Royal Society, 1699–1751." *Notes and Records of the Royal Society of London* 50 (1996): 13–27.

Arago, François. *Biographies of Distinguished Scientific Men by François Arago.* 2 vols. Translated by W. H. Smyth, Baden Powell, and Robert Grant. Boston: Ticknor and Fields, 1859.

Arnold, Thomas. *Observations on the Nature, Kinds, Causes, and Prevention of Insanity, Lunacy, or Madness.* 2 vols. Leicester, UK: G. Ireland, 1782–86.

Ash, Mitchell G. "Historicizing Mind Science: Discourse, Practice, Subjectivity." *Science in Context* 5 (1992): 193–207.

Ashworth, William J. "The Calculating Eye: Baily, Herschel, Babbage, and the Business of Astronomy." *British Journal for the History of Science* 27 (1994): 409–41.

———. "Memory, Efficiency, and Symbolic Analysis: Charles Babbage, John Herschel, and the Industrial Mind." *Isis* 87 (1996): 629–53.

———. "Memory, Foresight, and Production: The Work of Analysis in the Early Nineteenth Century." PhD thesis, University of Cambridge, 1996.

Babbage, Charles. *The Exposition of 1851; or, Views of the Industry, the Science, and the Government of England.* 2nd ed. London, 1851.

———. *On the Economy of Machinery and Manufactures.* 4th ed. 1835. Reprint, New York: Augustus M. Kelley, 1963.

———. *Passages from the Life of a Philosopher.* Edited by M. Campbell-Kelly. 1864. Reprint, New Brunswick, NJ: Rutgers University Press; Piscataway, NJ: IEEE Press, 1994.

———. *Reflections on the Decline of Science in England.* London: Fellowes, 1830.

———. *A Word to the Wise.* London: John Murray, 1833.

Bacon, Francis. *A History of Life and Death.* 1638. Extracts reprinted in Louis Cornaro, *The Art of Living Long.* Milwaukee: William F. Butler, 1903.

[Balfour, J. H.]. "Professor George Wilson: His Life and Writings." *North British Review* 32 (1860): 223–46.

Bancks's Manchester and Salford Directory: 1800. Manchester: G. Bancks, 1800.

Barker-Benfield, G. J. *The Culture of Sensibility: Sex and Society in Eighteenth-Century Britain.* Chicago: University of Chicago Press, 1992.

Barnes, E. J. "Fashioning a Natural Self: Guides to Self-Presentation in Victorian England." PhD thesis, University of Cambridge, 1995.

Barton, Ruth. "'An Influential Set of Chaps': The X-Club and Royal Society Politics 1864–85." *British Journal for the History of Science* 23 (1990): 53–81.

———. "'Men of Science': Language, Identity and Professionalization in the Mid-Victorian Scientific Community." *History of Science* 41 (2003): 73–119.

Bashford, Alison. *Purity and Pollution: Gender, Embodiment and Victorian Medicine.* Hampshire, UK: Macmillan; New York: St. Martin's Press, 1998.

Batten, Alan H. *Resolute and Undertaking Characters: The Lives of Wilhelm and Otto Struve.* Dordrecht: Reidel, 1988.

Baxter, Paul. "Brewster, Evangelicalism and the Disruption of the Church of Scotland." In Morrison-Low and Christie, *"Martyr of Science."*

Bell, Charles. *Idea of a New Anatomy of the Brain*. London: Strahan and Preston, 1811.

Berg, Maxine. *The Machinery Question and the Making of Political Economy, 1815-1848*. Cambridge: Cambridge University Press, 1980.

Berrios, German E. *The History of Mental Symptoms: Descriptive Psychopathology since the Nineteenth Century*. Cambridge: Cambridge University Press, 1996.

———, and Thomas R. Dening. "Pseudohallucinations: A Conceptual History." *Psychological Medicine* 26 (1996): 753-63.

Biagioli, Mario. *Galileo, Courtier: The Practice of Science in the Culture of Absolutism*. Chicago: University of Chicago Press, 1993.

———. "Tacit Knowledge, Courtliness, and the Scientist's Body." In *Choreographing History*, edited by Susan Leigh Foster. Bloomington: Indiana University Press, 1995.

Biot, Jean-Baptiste. *Life of Sir Isaac Newton*. Translated by Henry Brougham. London: Baldwin and Craddock, 1829.

Blakey, Robert. *History of the Philosophy of Mind; Embracing the Opinions of All Writers on Mental Science from the Earliest Period to the Present Time*, vol. 3. London: Longman, Brown, Green, and Longmans, 1850.

"Blunders of Vision—Color-Blindness." *Eclectic Magazine* 48 (1859): 513-17.

Bonnet, Charles. *Essai analytique sur les facultés de l'âme*. 2nd ed. Neuchatel: Fauche, 1769.

Bordo, Susan. *The Flight to Objectivity: Essays in Cartesianism and Culture*. Albany: State University of New York Press, 1987.

Boring, Edwin G. *Sensation and Perception in the History of Experimental Psychology*. New York: Appleton-Century-Crofts, 1942.

Bostock, John. *An Elementary System of Physiology*. London: Baldwin, Cradock, and Joy, 1824-27.

Boyle, Robert. *Experiments and Considerations Touching Colours*. Edited by Marie Boas Hall. 1664. Reprint, New York: Johnson Reprint Co., 1964.

———. "Some Uncommon Observations about Vitiated Sight." In *The Works of the Honourable Robert Boyle*, edited by Thomas Birch. New ed. London: J. and F. Rivington, 1772.

Brain, Robert M., and M. Norton Wise. "Muscles and Engines: Indicator Diagrams and Helmholtz's Graphical Methods." In *Universalgenie Helmholtz: Ruckblick nach 100 Jahren*, edited by Lorenz Krüger. Berlin: Akademie Verlag, 1994.

Brewer, John. *The Pleasures of the Imagination: English Culture in the Eighteenth Century*. London: Harper Collins, 1997.

[Brewster, David]. "Accidental Colours." *Edinburgh Encyclopaedia*. American ed. Philadelphia: Joseph and Edward Parker, 1832.

———. "Account of a Remarkable Case of Spectral Illusion, in Which Both the Eye and Ear Were Influenced." *Edinburgh Journal of Science* 2 (1830): 218-22, 319-21; 4 (1831): 261-63.

[———]."The Bridgewater Bequest; Whewell's *Astronomy and General Physics*." *Edinburgh Review* 58 (1834): 422-57.

[———]. "Dr. George Wilson *On Colour-Blindness*." *North British Review* 24 (1856): 325-58.

[———]. "Form and Colour." *North British Review* 32 (1860): 126-58.

[———]. "François Arago: His Life and Discoveries." *North British Review* 20 (1854): 459-500.

———. *Letters on Natural Magic, Addressed to Sir Walter Scott, Bart.* London: John Murray, 1832.

[———]. "Memoirs of John Dalton." *North British Review* 27 (1857): 465-97.

———. *Memoirs of the Life, Writings, and Discoveries of Sir Isaac Newton.* Edinburgh: Thomas Constable, 1855.

———. "Notice Respecting Certain Changes of Colour in the Choroid Coat of the Eye of Animals." *Philosophical Magazine* 3 (1833): 288-89.

———. "Observations on Colour Blindness, or Insensibility to the Impressions of Certain Colours." *Philosophical Magazine* 25 (1844): 134-41.

———. "On a New Property of the Retina." *Transactions of the Royal Society of Edinburgh* 24 (1865): 327-29.

———. "On Hemiopsy, or Half-Vision." *Transactions of the Royal Society of Edinburgh* 24 (1865): 15-18.

———. "On the Effect of Compression and Dilatation upon the Retina." *Philosophical Magazine* 1 (1832): 89-92.

———. "On the Optical Phenomena, Nature, and Locality of *Muscae Volitantes;* with Observations on the Structure of the Vitreous Humour, and on the Vision of Objects Placed within the Eye." *Transactions of the Royal Society of Edinburgh* 15 (1844): 377-85.

[———]. "Phrenology: Its Place and Relations." *North British Review* 17 (1852): 41-70.

[———]. "Works on Mental Philosophy, Mesmerism, Electro-biology, &c." *North British Review* 22 (1854): 179-224.

Brierre de Boismont, A.-J.-F. *Hallucinations; or, the Rational History of Apparitions, Visions, Dreams, Ecstasy, Magnetism, and Somnambulism.* 1853. Reprint, New York: Arno Press, 1976.

Brigham, Amariah. *Remarks on the Influence of Mental Cultivation and Mental Excitement upon Health.* 2nd ed. 1833. Reprint, Delmar, NY: Scholars' Facsimiles and Reprints, 1973.

Brock, W. H. "Brewster as a Scientific Journalist." In Morrison-Low and Christie, *"Martyr of Science."*

Brodie, Benjamin C. *Psychological Inquiries; in a Series of Essays Intended to Illustrate the Mutual Relations of the Physical Organization and the Mental Faculties.* 2nd ed. London: Longman, Brown, Green, and Longmans, 1855.

Brooke, John Hedley. *Science and Religion: Some Historical Perspectives.* Cambridge: Cambridge University Press, 1991.

———. "Scientific Thought and Its Meaning for Religion: The Impact of French Science on British Natural Theology, 1827-1859." *Revue de Synthèse* 110 (1989): 33-59.

Broussais, F.-J.-V. *Traité de physiologie appliquée à la pathologie.* 2 vols. Paris: Delauney, 1822-23.

[Brown, Samuel]. "Animal Magnetism." *North British Review* 15 (1851): 133-59.

Brown, Thomas. *Lectures on the Philosophy of the Human Mind.* 4 vols. Hallowell: Masters, Smith, and Co., 1850.

[———]. "Stewart's *Life of Dr. Reid*." *Edinburgh Review* 3 (1804): 269-87.

Browne, Janet. "I Could Have Retched All Night: Charles Darwin and His Body." In Lawrence and Shapin, *Science Incarnate.*

Browne, Thomas. *Pseudodoxica Epidemica.* Vol. 2 of *The Works of Thomas Browne.* Edited by Geoffrey Keene. Chicago: University of Chicago Press, 1964.

Buchwald, Jed Z. *The Rise of the Wave Theory of Light: Optical Theory and Experiment in the Early Nineteenth Century.* Chicago: University of Chicago Press, 1989.

———, ed. *Scientific Practice: Theories and Stories of Doing Physics.* Chicago: University of Chicago Press, 1995.

Bud, Robert, Barney Finn, and Helmuth Trischler, eds. *Manifesting Medicine: Bodies and Machines.* Amsterdam: Harwood Academic, 1999.

Buffon, Georges Louis Leclerc, Comte de. "Dissertation sur les couleurs accidentelles." *Mémoires de mathématique et de physique* 60 (1743): 147-58.

Burch, George J. "On Artificial Temporary Colour-Blindness, with an Examination of the Colour Sensations of 109 Persons." *Philosophical Transactions of the Royal Society of London,* ser. B, 191 (1899): 1-34.

Burke, John G. "Bursting Boilers and the Federal Power." In *Technology and Culture: An Anthology,* edited by Melvin Kranzberg and William H. Davenport. New York: Schocken Books, 1972.

Burney, Frances. *Madame D'Arblay's Diary and Letters.* London: Frederick Warne, 1890.

Burns, R. M. *The Great Debate on Miracles: From Joseph Glanvill to David Hume.* Lewisburg, PA: Bucknell University Press, 1981.

Burton, John Hill. *Life and Correspondence of David Hume.* . . . Edinburgh: W. Tait, 1846.

Busfield, Joan. "The Female Malady? Men, Women, and Madness in Nineteenth Century Britain." *Sociology* 28 (1994): 259-77.

Butter, John. "Remarks on the Insensibility of the Eye to Certain Colours." *Edinburgh Philosophical Journal* 6 (1822): 135–41.

Buttmann, Günther. *The Shadow of the Telescope: A Biography of John Herschel.* Translated by B. E. J. Pagel. New York: Charles Scribner's Sons, 1970.

Bynum, W. F. "The Nervous Patient in Eighteenth- and Nineteenth-Century Britain: The Psychiatric Origins of British Neurology." In Bynum, Porter, and Shepherd, *Anatomy of Madness.*

———, Roy Porter, and Michael Shepherd, eds. *Anatomy of Madness: Essays in the History of Psychiatry.* London: Tavistock, 1985.

Cabanis, P. J. G. *On the Relations between the Physical and Moral Aspects of Man.* 2 vols. Edited by G. Mora. Baltimore: Johns Hopkins University Press, 1981.

Cahan, David, ed. *From Natural Philosophy to the Sciences: Writing the History of Nineteenth-Century Science.* Chicago: University of Chicago Press, 2003.

Cain, Peter, ed. *Empire and Imperialism: The Debate of the 1870s.* Chicago: St. Augustine's Press, 2000.

Campbell, Lewis, and William Garnett. *The Life of James Clerk Maxwell; with a Selection from His Correspondence and Occasional Writings, and a Sketch of His Contributions to Science.* London: Macmillan, 1882.

Canguilhem, Georges. *The Normal and the Pathological.* Translated by C. R. Fawcett. 1966. Reprint, New York: Zone Books, 1991.

Cannon, Susan Faye. *Science in Culture: The Early Victorian Period.* New York: Dawson and Science History Publications, 1978.

——— [as Walter F.]. "John Herschel and the Idea of Science." *Journal of the History of Ideas* 22 (1961): 215–39.

Cantor, Geoffrey N. "Brewster on the Nature of Light." In Morrison-Low and Christie, *"Martyr of Science."*

———. "Henry Brougham and the Scottish Methodological Tradition." *Studies in History and Philosophy of Science* 2 (1971–72): 69–89.

———. "Light and Enlightenment: An Exploration of Mid-Eighteenth Century Modes of Discourse." In *The Discourse of Light from the Middle Ages to the Enlightenment.* Los Angeles: William Andrews Clark Memorial Library and University of California–Los Angeles, 1985.

———. *Michael Faraday: Sandemanian and Scientist: A Study of Science and Religion in the Nineteenth Century.* New York: St. Martin's Press, 1991.

———. *Optics after Newton: Theories of Light in Britain and Ireland, 1704–1840.* Manchester: Manchester University Press, 1983.

———. "The Theological Significance of Ethers." In Cantor and Hodge, *Conceptions of Ether.*

———, and M. J. S. Hodge, eds. *Conceptions of Ether: Studies in the History of Ether Theories 1740–1900*. Cambridge: Cambridge University Press, 1981.

Carbutt, Edward. "An Essay on the Signs of Ideas; or, the Means of Conveying to Others a Knowledge of Our Ideas." *Memoirs of the Manchester Literary and Philosophical Society* 3 (1819): 241–70.

Carlyle, Thomas. "Signs of the Times." In *A Carlyle Reader: Selections from the Writings of Thomas Carlyle*, edited by G. B. Tennyson. Cambridge: Cambridge University Press, 1984.

Carnot, Lazare. *Essai sur les machines en general*. 2nd ed. Dijon: Defay, 1786.

Carpenter, Elias. *Nocturnal Alarm; Being an Essay on Prophecy and Vision*. London: W. Smith, 1803.

Carpenter, William B. *Principles of Mental Physiology; With their Applications to the Training and Discipline of the Mind, and the Study of its Morbid Conditions*. 4th ed. New York: D. Appleton, 1877.

Carter, Margaret L. *Spectre or Delusion: The Supernatural in Gothic Fiction*. Ann Arbor, MI: UMI Research Press, 1987.

Castle, Terry. *The Female Thermometer: Eighteenth Century Culture and the Invention of the Uncanny*. Oxford: Oxford University Press, 1995.

———. "Phantasmagoria: Spectral Technology and the Metaphorics of Modern Reverie." *Critical Inquiry* 15 (1988): 26–61.

Cawood, John. "The Magnetic Crusade: Science and Politics in Early Victorian Britain." *Isis* 70 (1979): 493–518.

———. "Terrestrial Magnetism and the Development of International Collaboration in the Early Nineteenth Century." *Annals of Science* 34 (1977): 551–87.

Chadwick, Owen. *The Victorian Church*. London: Black, 1966–70.

Chalmers, Thomas. *On the Power, Wisdom, and Goodness of God as Manifested in the Adaptation of External Nature to the Moral and Intellectual Constitution of Man*. 2 vols. London, 1835.

Chambers, George F. *A Handbook of Descriptive and Practical Astronomy*. 4th ed. Oxford: Clarendon Press, 1890.

Chambers, Robert. *A Biographical Dictionary of Eminent Scotsmen*. 1870. Reprint, Hildesheim, NY: G. Olms, 1971.

Chapman, Allan. "Sir George Airy (1801–1892) and the Concept of International Standards in Science, Timekeeping, and Navigation." *Vistas in Astronomy* 28 (1985): 321–28.

———. "Private Research and Public Duty: George Biddell Airy and the Search for Neptune." *Journal for the History of Astronomy* 19 (1988): 121–39.

Chevreul, M. E. *De la loi du contraste simultané des couleurs, et de l'assortiment des objets colores, considere d'apres cette loi*. Paris: L. Laget, 1838.

Cheyne, George. *The English Malady*. London: G. Strahan and J. Leake, 1733.

———. *An Essay of Health and Long Life*. London: George Strahan, 1724.

Christie, J. R. R. "The Paracelsian Body." In *Paracelsus: The Man and His Reputation, His Ideas and Their Transformation*, edited by Ole Peter Grell. London: Brill, 1998.

Claparède, Edouard. *La psychologie animale de Charles Bonnet*. Geneva: Georg, 1909.

Clark, Stuart. "The Scientific Status of Demonology." In *Occult and Scientific Mentalities in the Renaissance*, edited by Brian Vickers. Cambridge: Cambridge University Press, 1984.

Clark, William, Jan Golinski, and Simon Schaffer, eds. *The Sciences in Enlightened Europe*. Chicago: University of Chicago Press, 1999.

Clarke, Edwin, and L. S. Jacyna. *Nineteenth-Century Origins of Neuroscientific Concepts*. Berkeley: University of California Press, 1987.

Clarke, Norma. "Strenuous Idleness: Thomas Carlyle and the Man of Letters as Hero." In *Manful Assertions: Masculinities in Britain since 1800*, edited by Michael Roper and John Tosh. London: Routledge, 1991.

Clarkson, Thomas. *A Portraiture of Quakerism*. . . . Indianapolis: Merrill and Field, 1870.

Clerke, Agnes Mary. *Modern Cosmogonies*. London: A. and C. Black, 1905.

Clery, E. J. *The Rise of Supernatural Fiction, 1762–1800*. Cambridge: Cambridge University Press, 1995.

Cody, John. "Watchers upon the East: The Ocular Complaints of Emily Dickinson." *Psychiatric Quarterly* 42 (1968): 548–76.

Coleridge, Samuel Taylor. *Aids to Reflection*. Edited by Henry Nelson Coleridge. New York: Stanford and Swords, 1854.

———. *The Friend; A Series of Essays to Aid in the Formation of Fixed Principles in Politics, Morals, and Religion, with Literary Amusements Interspersed*. Vol. 2 of *The Complete Works of Samuel Taylor Coleridge*. New York: Harper and Bros., 1856.

———. *Specimens of the Table Talk of Samuel Taylor Coleridge*. Edited by Alexander Whitson. New ed. London: John Murray, 1865.

Coleridge, Sara. *Memoirs and Letters of Sara Coleridge*. Edited by E. Coleridge. London: Henry S. King, 1873.

Colley, Linda. *Britons: Forging the Nation, 1707–1837*. New Haven: Yale University Press, 1992.

Colman, W. S. "Hallucinations in the Sane, Associated with Local Organic Disease of the Sensory Organs, etc." *British Medical Journal* 1 (1894): 1015–17.

Colp, Jr., Ralph. *To Be an Invalid: The Illness of Charles Darwin*. Chicago: University of Chicago Press, 1977.

Colquhoun, Hugh. "Account of Two Cases of Insensibility of the Eye to Certain of the Rays of Colour." *Glasgow Medical Journal* 2 (1829): 12–21.

Combe, Andrew. *Observations on Mental Derangement*. 1st American ed. 1834. Reprint, Delmar, NY: Scholars' Facsimiles and Reprints, 1972.

Combe, George. *Elements of Phrenology*. Edinburgh: Anderson; Simpkin and Marshall, 1824.

———. *A System of Phrenology*. 4th ed. Edinburgh: Maclachlan and Stewart, John Anderson, Jun., &c, 1836.

Conolly, John. *An Inquiry concerning the Indications of Insanity*. London: J. Taylor, 1830.

Cooke, William Fothergill. *Extracts from the Private Letters of the Late Sir William Fothergill Cooke, 1836–39, relating to the Invention and Development of the Electric Telegraph*. London: Spon, 1895.

Coon, Deborah J. "Standardizing the Subject: Experimental Psychologists, Introspection, and the Quest for a Technoscientific Ideal." *Technology and Culture* 34 (1993): 757–83.

Corsi, Pietro. *Science and Religion: Baden Powell and the Anglican Debate, 1800–1860*. Cambridge: Cambridge University Press, 1988.

Cottom, Daniel. *The Civilized Imagination: A Study of Ann Radcliffe, Jane Austen, and Sir Walter Scott*. Cambridge: Cambridge University Press, 1985.

Cranmore, R. T. "On Some Phaenomena of Defective Vision." *Philosophical Magazine* 36 (1850): 485–86.

Crary, Jonathan. *Techniques of the Observer: On Vision and Modernity in the Nineteenth Century*. Cambridge: MIT Press, 1990.

Crichton, Alexander. *An Inquiry into the Nature and Origin of Mental Derangement....* 2 vols. London: Cadell and Davies, 1798.

Crosland, Newton. *Apparitions: A New Theory*. 2nd ed. London: Effingham Wilson, 1856.

Crowe, Catherine. *The Night Side of Nature; or, Ghosts and Ghost Seers*. London: T. C. Newby, 1848.

Cullen, William. *Institutions of Medicine*. 3rd ed. Edinburgh: Charles Elliott; London: T. Cadell, 1785.

Cunningham, Andrew. "Getting the Game Right: Some Plain Words on the Identity and Invention of Science." *Studies in History and Philosophy of Science* 19 (1988): 365–89.

———. "How the *Principia* Got its Name; or, Taking Natural Philosophy Seriously." *History of Science* 29 (1991): 377–92.

———, and Perry Williams. "De-centring the Big Picture: *The Origins of Modern Science* and the Modern Origins of Science." *British Journal for the History of Science* 26 (1993): 407–32.

Dalton, John. *Elements of English Grammar*. Manchester, 1801.

——. "Extraordinary Facts Relating to the Vision of Colors, with Observations." *Memoirs and Proceedings of the Manchester Literary and Philosophical Society* 5 (1798): 28–45.

Danahay, Martin A. *A Community of One: Masculine Autobiography and Autonomy in Nineteenth-Century Britain*. Albany: State University of New York Press, 1993.

Danziger, Kurt. *Constructing the Subject: Historical Origins of Psychological Research*. Cambridge: Cambridge University Press, 1990.

Darwin, Charles. *The Autobiography of Charles Darwin 1809–1882*. Edited by Nora Barlow. New York: W. W. Norton and Company, 1958.

Darwin, Erasmus. *The Letters of Erasmus Darwin*. Edited by Desmond King-Hele. Cambridge: Cambridge University Press, 1981.

——. *Zoonomia; or, the Laws of Organic Life*. 3rd ed. London: J. Johnson, 1801.

Darwin, Robert Waring. "New Experiments on the Ocular Spectra of Light and Colours." *Philosophical Transactions of the Royal Society of London* 76 (1786): 313–48.

Daston, Lorraine. *Classical Probability in the Enlightenment*. Princeton: Princeton University Press, 1989.

——. "Fear and Loathing of the Imagination in Science." *Daedalus* 127, no. 1 (1998): 73–95.

——. "Knowledge of the Invisible, the Ineffable, and the Intuitive." Paper presented at the annual meeting of the History of Science Society. Atlanta, October 1994.

——. "Marvelous Facts and Miraculous Evidence in Early Modern Europe." *Critical Inquiry* 18 (1991): 93–124.

——. "Objectivity and the Escape from Perspective." *Social Studies of Science* 22 (1992): 597–618.

——, and Peter Galison. "The Image of Objectivity." *Representations* 40 (1992): 81–128.

——, and Katharine Park. *Wonders and the Order of Nature, 1150–1750*. New York: Zone Books, 2001.

Davy, Humphry. "Presidential Address on the Occasion of the Presentation of the First Royal Medal of the Royal Society to John Dalton." In vol. 7 of *The Collected Works of Sir Humphry Davy*, edited by John Davy. London: Smith, Elder, 1839.

——. *Researches, Chemical and Philosophical, Chiefly concerning Nitrous Oxide, or Dephlogisticated Nitrous Air, and Its Respiration*. London: Smith, Elder, 1839.

Davy, John. *Memoirs of the Life of Sir Humphry Davy, Bart.* 2 vols. London: Longman, Rees, Orme, Brown, Green, and Longman, 1836.

Dear, Peter. *Discipline and Experience: The Mathematical Way in the Scientific Revolution*. Chicago: University of Chicago Press, 1995.

——. "Miracles, Experiments, and the Ordinary Course of Nature." *Isis* 81 (1990): 663–83.

———. "Religion, Science and Natural Philosophy: Thoughts on Cunningham's Thesis." *Studies in History and Philosophy of Science* 32 (2001): 377–86.

Defoe, Daniel. *An Essay on the History and Reality of Apparitions.* London: J. Roberts, 1727.

"De Foe on Apparitions." *Blackwood's Edinburgh Magazine* 6 (1819): 201–7.

De La Hire, Philippe. "Dissertation sur les differens accidens de la vue." *Mémoires de l'Académie des Sciences de Paris; depuis 1666 jusqu'à 1699* 9 (1678): 530–634.

Dendy, Walter Cooper. *Philosophy of Mystery.* London: Longman, Orme, Brown, Green, and Longmans, 1841.

De Quincey, Thomas. *Confessions of an English Opium Eater.* Edited by Alethea Hayter. New York: Penguin, 1971.

Desmond, Adrian J. *The Politics of Evolution: Morphology, Medicine, and Reform in Radical London.* Chicago: University of Chicago Press, 1989.

Dickens, Charles. *Dombey and Son.* Edited by E. A. Horsman. Oxford: Clarendon, 1974.

Dickerson, Vanessa D. *Victorian Ghosts in the Noontide: Women Writers and the Supernatural.* Columbia: University of Missouri Press, 1996.

Dickinson, H. W. *Matthew Boulton.* Cambridge: Cambridge University Press, 1936.

Douard, John W. "E.-J. Marey's Visual Rhetoric and the Graphic Decomposition of the Body." *Studies in History and Philosophy of Science* 26 (1995): 175–204.

Drayton, Richard. *Nature's Government: Science, Imperial Britain, and the "Improvement" of the World.* New Haven: Yale University Press, 2000.

Duck, Michael. "Newton and Goethe on Colour: Physical and Physiological Considerations." *Annals of Science* 45 (1988): 507–19.

Duffin, Jacalyn M. "Sick Doctors: Bayle and Laennec on Their Own Phthisis." *Journal of the History of Medicine* 43 (1988): 165–82.

Duveen, D. I., and H. S. Klickstein. "John Dalton's Autopsy." *Journal of the History of Medicine* 9 (1954): 360–62.

Easton, David, and Corinne Schelling, eds. *Divided Knowledge: Across Disciplines, Across Cultures.* Newbury Park, CA: Sage, 1991.

Eastwood, David. *Governing Rural England: Tradition and Transformation in Local Government, 1780–1840.* Oxford: Clarendon Press, 1994.

Ebbecke, Ulrich. *Johannes Müller, der grosse rheinische Physiologe mit einem Neudruck von Johannes Müllers Schrift, "Über die phantastischen Gesichtserscheinungen."* Hanover: Schmorl und von Seefeld, 1951.

Edelcrantz, Abraham N. *A Treatise on Telegraphs; and Experiments with a New Construction Thereof.* 1796. Translated and reprinted in *The Early History of Data Networks,* by Gerard J. Holzmann and Björn Pehrson. Washington, DC: Wiley-IEEE Computer

Society, 1994. http://www.it.kth.se/docs/early_net/ch-1-4.html (accessed 20-21 May 1999).

Edgeworth, Maria. *Maria Edgeworth: Letters from England 1813-1844.* Edited by Christina Colvin. Oxford: Clarendon, 1971.

[Editorial Response]. *Gentleman's Magazine* 64 (September 1794): 815.

Elias, Norbert. *The Civilizing Process.* 2 vols. Basel: Haus zum Falker, 1939.

Elliot, John. *Elements of the Branches of Natural Philosophy Connected with Medicine.* London: J. Johnson, 1782.

Emerson, Roger L. "The Philosophical Society of Edinburgh 1768-1783." *British Journal for the History of Science* 18 (1985): 255-303.

———. "Science and the Origins and Concerns of the Scottish Enlightenment." *History of Science* 26 (1988): 333-66.

———. "The Scottish Enlightenment and the End of the Philosophical Society of Edinburgh." *British Journal for the History of Science* 21 (1988): 33-66.

Emmons, S. Bulfinch. *Philosophy of Popular Superstitions.* Boston: L. P. Crown, 1853.

Esquirol, Etienne. *Mental Maladies: A Treatise on Insanity.* Translated by E. K. Hunt. 1845. Reprint, New York: Hafner, 1965.

Everitt, C. W. F. *James Clerk Maxwell: Physicist and Natural Philosopher.* New York: Charles Scribner's Sons, 1975.

Fahie, J. J. *A History of Electric Telegraphy, to the Year 1837.* London: E. and F. N. Spon, 1884.

Faraday, Michael. "Observations on the Education of the Judgment." In *Modern Culture; Its True Aims and Requirements: A Series of Addresses and Arguments on the Claims of Scientific Education,* edited by Edward L. Youmans. London: Macmillan, 1867.

———. *The Selected Correspondence of Michael Faraday.* Edited by L. Pearce Williams. Cambridge: Cambridge University Press, 1971.

Farrar, W. V., K. Farrar, and E. L. Scott. "The Henrys of Manchester, Part 2; Thomas Henry's Sons: Thomas, Peter, and William." *Ambix* 21 (1974): 179-207.

Farrington, Benjamin. "*Temporis Partus Masculus:* An Untranslated Writing of Francis Bacon." *Centaurus* 1 (1951): 193-205.

Fechner, Gustav. *Nanna; oder über das Seelenleben der Pflanzen.* Leipzig: L. Voss, 1848.

———. *Zend-Avesta; oder über die Dinge des Himmels und des Jenseits.* 3 vols. Leipzig: L. Voss, 1851.

Feher, Michel, ed. *Fragments for a History of the Human Body.* 3 vols. Cambridge: MIT Press, 1989.

Ferriar, John. *An Essay toward a Theory of Apparitions.* London: Cadell and Davies, 1813.

———. "Of Popular Illusions, and Particularly of Medical Demonology." *Memoirs and Proceedings of the Manchester Literary and Philosophical Society* 3 (1790): 31–116.

"Ferriar on *Apparitions.*" *Quarterly Review* 9 (1813): 304–12.

Figlio, Karl M. "Theories of Perception and the Physiology of Mind in the Late Eighteenth Century." *History of Science* 13 (1975): 177–212.

Fissell, Mary. "The Disappearance of the Patient's Narrative and the Invention of Hospital Medicine." In French and Wear, *British Medicine.*

Flint, Kate. *The Victorians and the Visual Imagination.* Cambridge: Cambridge University Press, 2000.

Flynn, Carol Houlihan. "Running Out of Matter: The Body Exercised in Eighteenth-Century Fiction." In Rousseau and Porter, *Languages of Psyche.*

Forbes, J. D. *The Danger of Superficial Knowledge.* London: J. W. Parker, 1849.

Forrest, D. W. *Francis Galton: The Life and Work of a Victorian Genius.* London: Paul Elek, 1974.

Forster, Thomas. "On a Systematic Arrangement of Colours." *Philosophical Magazine* 42 (1813): 119–21, 327–29.

Fothergill, John. "Remarks on the Sick Headach." *Medical Observations and Inquiries* 6 (1784): 103–37.

Foucault, Michel. *The Birth of the Clinic: An Archaeology of Medical Perception.* Translated by A. M. Sheridan Smith. New York: Pantheon Books, 1973.

Frank, Arthur W. "Bringing Bodies Back In: A Decade Review." *Theory, Culture, and Society* 7 (1990): 131–62.

Franklin, Benjamin. *The Autobiography of Benjamin Franklin.* Edited by Max Farrand. Berkeley: University of California Press, 1949.

French, R. K. *Robert Whytt, the Soul, and Medicine.* London: Wellcome Institute for the History of Medicine, 1969.

———, and Andrew Wear, eds. *British Medicine in the Age of Reform.* London: Routledge, 1991.

Fyfe, Aileen. "The Reception of William Paley's *Natural Theology* in the University of Cambridge." *British Journal for the History of Science* 30 (1997): 321–35.

Gabbey, Alan. "Newton and Natural Philosophy." In *The Companion to the History of Modern Science,* edited by Robert Olby et al. London: Routledge, 1990.

Galison, Peter. *How Experiments End.* Chicago: University of Chicago Press, 1987.

Gall, Franz Josef, and G. Spurzheim. *Anatomie et physiologie du système nerveux.* Paris: F. Schoell, 1810–19.

Galton, Francis. *Inquiries into Human Faculty and Its Development.* 2nd ed. London: J. M. Dent, 1892.

———. *Memories of My Life*. London: Methuen, 1909.

———. "Mental Imagery." *Popular Science Monthly* 18 (1880–81): 64–76.

———. "Statistics of Mental Imagery." *Mind* 5 (1880): 301–18.

———. "The Visions of Sane Persons." *Popular Science Monthly* 19 (1881): 519–31.

Gamble, John. *An Essay on the Different Modes of Communication with Signals....* London: W. Miller, 1797.

Gascoigne, John. *Cambridge in the Age of Enlightenment: Science, Religion, and Politics from the Restoration to the French Revolution*. Cambridge: Cambridge University Press, 1989.

———. "From Bentley to the Victorians: The Rise and Fall of British Newtonian Natural Theology." *Science in Context* 2 (1988): 219–56.

———. *Joseph Banks and the English Enlightenment: Useful Knowledge and Polite Culture*. Cambridge: Cambridge University Press, 1994.

Geikie, Archibald. *Life of Sir Roderick I. Murchison*. London: John Murray, 1875.

Geison, Gerald L. "Social and Institutional Factors in the Stagnancy of English Physiology, 1840–1870." *Bulletin of the History of Medicine* 46 (1972): 30–58.

Gieryn, Thomas F. "Boundaries of Science." In *Handbook of Science and Technology Studies*, edited by Sheila Jasanoff et al. Newbury Park, CA: Sage, 1994.

Gillespie, Neal C. "Divine Design and the Industrial Revolution: William Paley's Abortive Reform of Natural Theology." *Isis* 81 (1990): 214–29.

Gillispie, Charles Coulston, ed. *Dictionary of Scientific Biography*. New York: Scribner, 1970–90.

Gladstone, J. H. "Sir David Brewster." *Proceedings of the Royal Society of London* 17 (1868–69): xix–xxiv.

Gladstone, William E. "The Colour-Sense." *The Nineteenth Century* 2 (1877): 366–88.

———. *Studies on Homer and the Homeric Age*. Oxford: Oxford University Press, 1858.

Glanvill, Joseph. *Saducismus triumphatus; or, Full and Plain Evidence concerning Witches and Apparitions*. Edited by Henry More. 1681. Reprint, Gainesville, FL: Scholars' Facsimiles and Reprints, 1966.

Gleason, Mary Louise. *The Royal Society of London: Years of Reform, 1827–1847*. New York: Garland Publishing, 1991.

Goethe, Johann Wolfgang von. *Goethe's sammtliche Werke in dreitzig Banden*. 30 vols. Stuttgart: Cotta, 1850–51.

———. *Theory of Colours*. Translated by Charles Eastlake. 1840. Reprint, Cambridge: MIT Press, 1970.

———. *Zur Farbenlehre*. Vol. 13 of *Goethes Werke*. Hamburg: Christian Wagner Verlag, 1960.

——. *Zur Naturwissenschaft uberhaupt, besonders zur Morphologie: Erfahrung, Betrachtung, Folgerung, durch Lebensereignisse verbunden.* Stuttgart: J. G. Cotta, 1800.

Golinski, Jan. *Science as Public Culture: Chemistry and Enlightenment in Britain, 1760–1820.* Cambridge: Cambridge University Press, 1992.

Good, Gregory. "John Herschel's Optical Researches and the Development of His Ideas on Method and Causality." *Studies in History and Philosophy of Science* 18 (1987): 1–41.

Gooding, David. "'He Who Proves, Discovers': John Herschel, William Pepys and the Faraday Effect." *Notes and Records of the Royal Society of London* 39 (1985): 229–44.

——. "How Do Scientists Reach Agreement about Novel Observations?" *Studies in History and Philosophy of Science* 17 (1986): 205–30.

——. "Metaphysics versus Measurement: The Conversion and Conservation of Force in Faraday's Physics." *Annals of Science* 37 (1980): 1–29.

Gordon, Margaret Maria. *The Home Life of Sir David Brewster.* Edinburgh: Edmonton and Douglas, 1869.

Graves, Robert. *Life of Sir William Rowan Hamilton.* 1882. Reprint, New York: Arno Press, 1975.

Greenblatt, Stephen. *Renaissance Self-Fashioning: From More to Shakespeare.* Chicago: University of Chicago Press, 1980.

Green Musselman, Elizabeth. "Katherine Clerk Maxwell." *The Dictionary of Nineteenth-Century British Scientists.* Edited by Bernard Lightman. Bristol, UK: Thoemmes Continuum, 2004.

——. "Worlds Displaced: Projecting the Celestial Environment from the Cape Colony," *Kronos: Journal of Cape History* 29 (2003): 64–85.

Griffin, D. "On an Unusual Affection of the Eye, in Which Three Images Were Produced." *Philosophical Magazine* 6 (1835): 281–84.

Guerrini, Anita. "Case History as Spiritual Autobiography: George Cheyne's 'Case of the Author.'" *Eighteenth-Century Life* 19 (1995): 18–27.

Gurney, Edmund, Frederic W. H. Myers, and Frank Podmore. *Phantasms of the Living.* London: Society for Psychical Research, 1886.

Hacking, Ian. *The Taming of Chance.* Cambridge: Cambridge University Press, 1990.

Hahn, Roger. "Laplace and the Mechanistic Universe." In *God and Nature: Historical Essays on the Encounter between Christianity and Science,* edited by David C. Lindberg and Ronald L. Numbers. Berkeley: University of California Press, 1986.

Hakfoort, Casper. *Optics in the Age of Euler: Conceptions of the Nature of Light, 1700–1795.* Cambridge: Cambridge University Press, 1995.

Haley, Bruce. *The Healthy Body and Victorian Culture.* Cambridge: Harvard University Press, 1978.

Hamilton, William. *Discussions on Philosophy*. 2nd ed. London: Longman, Brown, Green and Longman, 1853.

Hamlin, Christopher. *Public Health and Social Justice in the Age of Chadwick: Britain, 1800–1854*. Cambridge: Cambridge University Press, 1998.

Hankins, Thomas L., and Robert J. Silverman. *Instruments and the Imagination*. Princeton: Princeton University Press, 1995.

Hannaway, Owen. *The Chemists and the Word: The Didactic Origins of Chemistry*. Baltimore: Johns Hopkins University Press, 1975.

Hargreave, David. "Thomas Young's Theory of Color Vision: Its Roots, Development, and Acceptance by the British Scientific Community." PhD diss., University of Wisconsin, 1973.

Harrison, J. A. "'Blind Henry Moyes': An Excellent Lecturer in Philosophy." *Annals of Science* 13 (1957): 109–25.

Hartley, David. *Observations on Man, His Frame, His Duty, and His Expectations*. Hildesheim: George Olms, 1967.

Harvey, George. "On an Anomalous Case of Vision with Regard to Colours." *Transactions of the Royal Society of Edinburgh* 10 (1824): 253–62.

Hatfield, Gary. "Remaking the Science of Mind: Psychology as Natural Science." In *Inventing Human Science: Eighteenth Century Domains*, edited by Christopher Fox, Roy Porter, and Robert Wokler. Berkeley: University of California Press, 1995.

Hawes, Clement. *Mania and Literary Style: The Rhetoric of Enthusiasm from the Ranters to Christopher Smart*. Cambridge: Cambridge University Press, 1996.

Hawkins, Anne Hunsaker. *Reconstructing Illness: Studies in Pathography*. West Lafayette, IN: Purdue University Press, 1993.

Hawksworth, Hallam [Francis Blake Atkinson]. *The Workshop of the Mind*. New York: Century, 1923.

Haüy, René-Just. *Traité élémentaire de physique*. 2nd ed. Paris: Chez Courcier, 1803.

Hay, David Ramsay. *A Nomenclature of Colours, Hues, Tints, and Shades, Applicable to the Arts and Natural Sciences, to Manufactures, and Other Purposes of General Utility*. Edinburgh: W. Blackwood, 1845.

———. *The Principles of Beauty in Colouring Systematized*. Edinburgh: William Blackwood and Sons, 1845.

Hayter, Alethea. *Opium and the Romantic Imagination*. Berkeley: University of California Press, 1970.

Heimann, P. M. "The Unseen Universe: Physics and the Philosophy of Nature in Victorian Britain." *British Journal for the History of Science* 6 (1972): 73–79.

———. "Voluntarism and Immanence: Conceptions of Nature in Eighteenth-Century Thought." *Journal of the History of Ideas* 39 (1978): 271–83.

Heineken, N. S. "On a Singular Irregularity of Vision." *Philosophical Magazine* 33 (1848): 318.

Helling, Georg L. A. *Praktisches Handbuch der Augenkrankheiten.* Berlin: F. Dümmler, 1821-22.

Helmholtz, Hermann von. *Helmholtz's Treatise on Physiological Optics.* Edited and translated by James P. C. Southall. New York: Dover, 1962.

———. *Popular Lectures on Scientific Subjects.* Translated by E. Atkinson. New York: D. Appleton, 1885.

———. "Über die Erhaltung der Kraft: Eine physikalische Abhandlung." In vol. 1 of *Wissenschaftliche Abhandlungen.* Leipzig: Barth, 1882-95.

Helmstadter, Richard J., and Bernard Lightman, eds. *Victorian Faith in Crisis: Essays on Continuity and Change in Nineteenth-Century Religious Belief.* Stanford: Stanford University Press, 1991.

Hempton, David. *Religion and Political Culture in Britain and Ireland: From the Glorious Revolution to the Decline of Empire.* Cambridge: Cambridge University Press, 1996.

Henry, John. "Occult Qualities and the Experimental Philosophy: Active Principles in Pre-Newtonian Matter Theory." *History of Science* 24 (1986): 335-81.

Henry, William. *The Elements of Experimental Chemistry.* 11th ed. London: Baldwin and Cradock, 1829.

———. "An Estimate of the Philosophical Character of Dr. Priestley." *Philosophical Magazine* 11 (1832): 207-18.

Henry, William Charles. *Memoirs of the Life and Scientific Researches of John Dalton.* London: Cavendish Society, 1848.

Hermann, Dieter B. *The History of Astronomy from Herschel to Hertzsprung.* Cambridge: Cambridge University Press, 1984.

Herschel, John. "1845 Presidential Address [to the British Association for the Advancement of Science]." In *Essays from the Edinburgh and Quarterly Reviews.*

———. *Essays from the Edinburgh and Quarterly Reviews, with Addresses from Other Pieces.* London: Longmans, 1857.

———. "Light." In vol. 4 of the *Encyclopaedia Metropolitana,* edited by Edward Smedley, Hugh James Rose, and Henry John Rose. London: B. Fellowes, 1845.

———. "Man, the Interpreter of Nature." In *Essays from the Edinburgh and Quarterly Reviews.*

———. "Mathematics." In vol. 13 of the *Edinburgh Encyclopaedia.* Edinburgh: W. Blackwood, 1830.

[———]. "*Mechanism of the Heavens,* by Mrs. Somerville." *Quarterly Review* 47 (1832): 537-59.

"On the Action of Crystallized Bodies on Homogeneous Light, and on the Causes of the Deviation from Newton's Scale in the Tints Which Many of Them Develope on

Exposure to a Polarised Ray." *Philosophical Transactions of the Royal Society of London* 110 (1820): 45–100.

———. "On Sensorial Vision." In *Familiar Lectures on Scientific Subjects*. London: A. Strahan, 1867.

———. *A Preliminary Discourse on the Study of Natural Philosophy*. 1830. Reprint, Chicago: University of Chicago Press, 1987.

[———]. "Whewell on the Inductive Sciences." *Quarterly Review* 68 (1841): 177–238.

———, and James South. "Observations of the Apparent Distances and Positions of 380 Double and Triple Stars." *Philosophical Transactions of the Royal Society of London* 114 (1824): 1–412.

Herschel, William. "Investigation of the Cause of That Indistinctness of Vision Which Has Been Ascribed to the Smallness of the Optic Pencil." *Philosophical Transactions of the Royal Society of London* 76 (1786): 500–7.

Hevly, Bruce. "The Heroic Science of Glacier Motion." *Osiris* 11 (1996): 66–86.

Heyck, T. W. "From Men of Letters to Intellectuals: The Transformation of Intellectual Life in Nineteenth-Century England." *Journal of British Studies* 20 (1980): 158–83.

Hibbert, Samuel. *Sketches of the Philosophy of Apparitions; or, An Attempt to Trace Such Illusions to Their Physical Causes*. 1824. Reprint, New York: Arno Press, 1975.

Hill, Brian. "A Life in the Laboratory: William Hyde Wollaston, M.D., F.R.S." *Practitioner* 197 (1966): 235–40.

Hogg, Jabez. "Colour-Blindness." *Popular Science Review* 2 (1863): 497–509.

Hole, Robert. *Pulpits, Politics, and Public Order in England 1760–1832*. Cambridge: Cambridge University Press, 1989.

Holland, Henry. *Chapters on Mental Physiology*. London: Longman, Brown, Green, and Longmans, 1852.

[———]. "Life of Dalton." *Quarterly Review* 96 (1854): 43–75.

Holmes, Tom W. *The Semaphore: The Story of the Admiralty-to-Portsmouth Shutter Telegraph and Semaphore Lines 1796 to 1847*. Ilfracombe, Devon, UK: Arthur H. Stockwell, 1983.

Holzmann, Gerard J., and Björn Pehrson. *The Early History of Data Networks*. Washington, DC: Wiley-IEEE Computer Society, 1994. http://www.it.kth.se/docs/early_net/toc.html (accessed 20–21 May 1999).

Hombres-Firmas, Louis Augustin d'. "Nouvelles observations d'achromatopsie." *Comptes Rendus* 30 (1850): 56–60, 376–79.

———. "Observations d'achromatopsie." *Comptes Rendus* 29 (1849): 175–79.

Hoover, Suzanne R. "Coleridge, Humphry Davy, and Some Early Experiments with a Consciousness-Altering Drug." *Bulletin of Research in the Humanities* 81 (1978): 9–27.

Horton, Susan R. "Were They Having Fun Yet? Victorian Optical Gadgetry, Modernist Selves." In *Victorian Literature and the Victorian Visual Imagination*, edited by Carol T. Christ and John O. Jordan. Berkeley: University of California Press, 1995.

Huddart, Joseph. "An Account of Persons Who Could Not Distinguish Colours." *Philosophical Transactions of the Royal Society of London* 67 (1777): 260-65.

Huggins, Margaret L. M. *Agnes Mary Clerke and Ellen Mary Clerke: An Appreciation*. Private publication, 1907.

Hume, David. *Essays: Moral, Political, and Literary*. Vol. 4 of *The Philosophical Works*. Edited by T. H. Green and T. H. Grose. Aalen: Scientia, 1964.

——. *History of England from the Invasion of Julius Caesar to the Revolution in 1688*. New ed. London, 1822.

Hunt, Bruce J. "The Ohm Is Where the Art Is: British Telegraph Engineers and the Development of Electrical Standards." *Osiris* 9 (1994): 48-63.

Hunt, David M., Kanwaljit S. Dulai, James K. Bowmaker, and John D. Mollon. "The Chemistry of John Dalton's Color Blindness." *Science* 267 (1995): 984-88.

Hunter, Kathryn Montgomery. *Doctor's Stories: The Narrative Structure of Medical Knowledge*. Princeton: Princeton University Press, 1991.

Hutchison, Keith. "Idiosyncrasy, Achromatic Lenses, and Early Romanticism." *Centaurus* 34 (1991): 125-71.

Huxley, T. H. "A Liberal Education, and Where to Find It." In *Science and Education*. New York: D. Appleton, 1896.

Hyman, Anthony. *Charles Babbage: Pioneer of the Computer*. Princeton: Princeton University Press, 1983.

Iliffe, Rob. "Isaac Newton: Lucatello Professor of Mathematics." In Lawrence and Shapin, *Science Incarnate*.

Inglis, D. "Hallucinations and Illusions of the Sane." *American Lancet* 18 (1894): 81-83.

Ingram, John H. *The Haunted Homes and Family Traditions of Great Britain*. London: Reeves and Turner, 1812.

Inkster, Ian. "London Science and the Seditious Meetings Act of 1817." *British Journal for the History of Science* 12 (1979): 192-96.

——. *Science and Technology in History: An Approach to Industrial Development*. New Brunswick, NJ: Rutgers University Press, 1991.

——. "The Social Context of an Educational Movement: A Revisionist Approach to the English Mechanics' Institutes, 1820-1850." In *Scientific Culture and Urbanisation in Industrialising Britain*. Brookfield, VT: Variorum, 1997.

——, and Jack Morrell, eds. *Metropolis and Province: Science in British Culture, 1780-1850*. Philadelphia: University of Pennsylvania Press, 1983.

Jackson, J. Hughlings. "Commentary on a Case of Hemiplegia with Amaurosis, with General Remarks on Amaurosis from Disease of the Brain." *The Ophthalmic Review* 3 (April 1866): 42–53.

Jackson, Myles W. "Artisanal Knowledge and Experimental Natural Philosophers: The British Response to Joseph Fraunhofer and the Bavarian Usurpation of Their Optical Empire." *Studies in History and Philosophy of Science* 25 (1994): 549–75.

———. "A Spectrum of Belief: Goethe's 'Republic' versus Newtonian 'Despotism.'" *Social Studies of Science* 24 (1994): 673–701.

Jacyna, L. S. *Philosophic Whigs: Medicine, Science, and Citizenship in Edinburgh, 1789–1848.* New York: Routledge, 1994.

Jardine, Nicholas. "Inner History; or, How to End Enlightenment." In Clark et al., *The Sciences in Enlightened Europe.*

Jeffries, B. Joy. *Color-Blindness: Its Dangers and Its Detection.* Boston: Houghton, Osgood, 1879.

Jennings, Humphrey, Mary-Lou Jennings, and Charles Madge. *Pandaemonium 1660–1886: The Coming of the Machine as Seen by Contemporary Observers.* London: Papermac, 1995.

Jones, Colin, and Roy Porter, eds. *Reassessing Foucault: Power, Medicine, and the Body.* London: Routledge, 1994.

Jordan, J. L. "Observations on a Singular Phenomenon Called the Spectre of the Broken." *Philosophical Magazine* 1 (1798): 232–35.

Jurin, James. "An Essay upon Distinct and Indistinct Vision." In *A Compleat System of Opticks in Four Books,* by Robert Smith, 115–71. Cambridge, 1738.

Kapur, Narinder. *Injured Brains of Medical Minds: Views from Within.* Oxford: Oxford University Press, 1997.

[Kater, Henry]. "An Account of Some Curious Facts Respecting Vision." *Philosophical Magazine* 5 (1834): 375–76.

Keller, Evelyn Fox. *Reflections on Gender and Science.* New Haven: Yale University Press, 1985.

Ketabgian, Tamara. "The Human Prosthesis: Workers and Machines in the Victorian Industrial Scene." *Critical Matrix* 11 (1997): 4–32.

King, Peter. *The Life of John Locke, with Extracts from his Correspondence, Journals, and Common-place Books.* 2nd ed. London: H. Colburn and R. Bentley, 1830.

King-Hele, Desmond. *Doctor of Revolution: The Life and Genius of Erasmus Darwin.* London: Faber and Faber, 1977.

Kitteringham, Guy. "Science in Provincial Society: The Case of Liverpool in the Early Nineteenth Century." *Annals of Science* 39 (1982): 329–48.

Knox, Kevin C. "Dephlogisticating the Bible: Natural Philosophy and Religious Controversy in Late Georgian Cambridge." *History of Science* 34 (1996): 167–200.

Kremer, Richard L. *The Thermodynamics of Life and Experimental Physiology, 1770-1880.* New York: Garland, 1990.

Krüger, Lorenz, et al., eds. *The Probabilistic Revolution.* 2 vols. Cambridge: Harvard University Press, 1987.

Kuklick, Henrika. "The Color Blue: From Research in the Torres Strait to an Ecology of Human Behaviour." In *Darwin's Laboratory: Evolutionary Theory and Natural History in the Pacific,* edited by Roy MacLeod and P. F. Rehbock. Honolulu: University of Hawai'i Press, 1994.

Kusch, Martin. *Psychological Knowledge: A Social History and Philosophy.* London: Routledge, 1999.

Ladd-Franklin, Christine. *Colour and Colour Theories.* New York: Harcourt, Brace and Co.; London: Kegan, Paul, 1929.

La Mettrie, Julien Offray de. *L'Homme Machine.* 1748. Reprint, La Salle, IL: Open Court, 1987.

Lang, Andrew. *Cock Lane and Common-Sense.* New York: AMS Press, 1970.

Langford, Paul. *Public Life and the Propertied Englishman 1689-1798.* Oxford: Clarendon, 1991.

Laqueur, Thomas. *Making Sex: Body and Gender from the Greeks to Freud.* Cambridge: Harvard University Press, 1990.

Larmor, Joseph, ed. *Memoir and Scientific Correspondence of the Late Sir George Gabriel Stokes.* Cambridge: Cambridge University Press, 1907.

"Last Case of Colour-Blindness." *Punch* 49 (1865): 216.

Laudan, Larry. *Science and Hypothesis: Historical Essays on Scientific Methodology.* Dordrecht: D. Reidel, 1981.

Laudan, Rachel. "The Role of Methodology in Lyell's Science." *Studies in History and Philosophy of Science* 13 (1982): 215-49.

Lavater, Ludwig. *Of Ghostes and Spirites Walking by Night and of Straunge Noyses, Crackes, and Sundrie Forewarnings, Which Commonly Happen before the Death of Men.* Translated by Robert Harrison. London: T. Creede, 1596.

Lawrence, Christopher. "The Nervous System and Society in the Scottish Enlightenment." In *Natural Order: Historical Studies of Scientific Culture,* edited by Barry Barnes and Steven Shapin. Beverly Hills, CA: Sage, 1979.

———, and Steven Shapin, eds. *Science Incarnate: Historical Embodiments of Natural Knowledge.* Chicago: University of Chicago Press, 1998.

Lenoir, Timothy. "Helmholtz and the Materialities of Communication." *Osiris* 9 (1994): 185-207.

Levene, John R. "Sir G. B. Airy, F.R.S. (1801-1892) and the Symptomatology of Migraine." *Notes and Records of the Royal Society of London* 30 (1975): 15-23.

——. "Sir George Biddell Airy, F.R.S. (1801–1892) and the Discovery and Correction of Astigmatism." *Notes and Records of the Royal Society of London* 21 (1966): 180–99.

Levêsque de Pouilly, Jean Simon. *Éloge de Charles Bonnet*. Lausanne: Henbach, 1794.

[Lewes, George Henry]. "Great Wits, Mad Wits?" *Blackwood's Magazine* 88 (1860): 302–11.

——. *The Physical Basis of the Mind*. Boston: James R. Osgood, 1877.

Lindberg, David C. *Theories of Vision from Al-kindi to Kepler*. Chicago: University of Chicago Press, 1976.

Lloyd, Genevieve. *The Man of Reason: "Male" and "Female" in Western Philosophy*. Minneapolis: University of Minnesota Press, 1984.

Locke, John. *An Essay concerning Human Understanding*. Edited by Peter H. Nidditch. 1689. Reprint, Oxford: Clarendon Press, 1975.

Lockhart, J. G. *Memoirs of Sir Walter Scott*. Edinburgh: Adam and Charles Black, 1869.

Logan, Peter Melville. *Nerves and Narratives: A Cultural History of Hysteria in Nineteenth-Century British Prose*. Berkeley: University of California Press, 1997.

Lonsdale, Henry. *John Dalton*. Vol. 5 of *The Worthies of Cumberland*. London: Routledge, 1874.

Lowe, Philip. "The British Association and the Provincial Public." In MacLeod and Collins, *Parliament of Science*.

MacKenzie, James. *The History of Health and the Art of Preserving It*. Edinburgh, 1758.

MacLeod, Roy, and Peter Collins, eds. *The Parliament of Science: The British Association for the Advancement of Science 1831–1981*. Northwood, Middlesex, UK: Science Reviews, 1981.

Maehle, Andreas-Holger. "Pharmacological Experimentation with Opium in the Eighteenth Century." In *Drugs and Narcotics in History*, edited by Roy Porter and Mikuláš Teich. Cambridge: Cambridge University Press, 1995.

Magendie, François. *Précis Elémentaire de Physiologie*. 2 vols. Paris: Mequignon-Marvis, 1816–17.

Marey, E. J. *La Méthode graphique dans les sciences expérimentales et particulièrement en physiologie et en médecine*. Paris, 1878.

Marillier, Leon. "Statistique des hallucinations." *Congrès Internationale de Psychologie*. Paris, 1890.

Mason, Michael. *The Making of Victorian Sexuality*. Oxford: Oxford University Press, 1995.

Mathias, Peter. *The First Industrial Revolution: An Economic History of Britain, 1700–1914*. London: Methuen, 1983.

Matthews, J. Rosser. *Quantification and the Quest for Medical Certainty*. Princeton: Princeton University Press, 1995.

Maudsley, Henry. *Natural Causes and Supernatural Seemings*. London: Kegan Paul, 1886.

Maxwell, James Clerk. *The Scientific Letters and Papers of James Clerk Maxwell*. Edited by Peter M. Harman. 2 vols. Cambridge: Cambridge University Press, 1990.

———. *Scientific Papers of James Clerk Maxwell*. Edited by W. D. Niven. 2 vols. Paris: Libraire Scientifique J. Hermann, 1927.

McConnell, Anita. *Instrument Makers to the World: A History of Cooke, Troughton and Simms*. York: William Sessions, 1992.

McDermot, Martin. *A Critical Dissertation on the Nature and Principles of Taste*. London: Sherwood, Jones, and Co., 1823.

McGucken, William. *Nineteenth Century Spectroscopy*. Baltimore: Johns Hopkins University Press, 1969.

McGuire, J. E., and Martin Tamny. *Certain Philosophical Questions: Newton's Trinity Notebook*. Cambridge: Cambridge University Press, 1986.

McNeil, Maureen. *Under the Banner of Science: Erasmus Darwin and His Age*. Manchester: Manchester University Press, 1987.

Meadows, A. J. *Recent History (1836-1975)*. Vol. 2 of *Greenwich Observatory: The Royal Observatory at Greenwich and Herstmonceux 1675-1975*. London: Taylor and Francis, 1975.

Merz, J. T. *A History of European Thought in the Nineteenth Century*. 4 vols. Edinburgh: William Blackwood and Sons, 1923-50.

Mialet, Hélène. "De-construction of a Genius: The Fabrication of Stephen Hawking." Paper presented at the Conférence Bielefeld, October 1996.

Micale, Mark S. *Approaching Hysteria: Disease and Its Interpretations*. Princeton: Princeton University Press, 1995.

Mill, John Stuart. *Autobiography of John Stuart Mill*. 1873. Reprint, New York: Columbia University Press, 1924.

———. *The Letters of John Stuart Mill*. Edited by Hugh S. R. Elliot. London: Longmans, Green, and Co., 1910.

Miller, David Philip. "Between Hostile Camps: Sir Humphry Davy's Presidency of the Royal Society of London, 1820-1827." *British Journal for the History of Science* 16 (1983): 1-47.

———. "The Royal Society of London 1800-1835: A Study in the Cultural Politics of Scientific Organization." PhD diss., University of Pennsylvania, 1981.

Millington, J. P. *John Dalton*. London: J. M. Dent; New York: E. P. Dutton, 1906.

Mitchell, Maria. "Maria Mitchell's Reminiscences of the Herschels." *Century Illustrated Monthly Magazine* 38 (1889): 903-9.

Moigno, Abbé François Napoleon Marie. *Repertoire d'optique moderne*. Paris: A. Franck, 1847.

Morrell, Jack. "Economic and Ornamental Geology: The Geological and Polytechnic Society of the West Riding of Yorkshire, 1837-53." In Inkster and Morrell, *Metropolis and Province.*

———. Individualism and the Structure of British Science in 1830." *Historical Studies in the Physical Sciences* 3 (1971): 183-204.

———. "Reflections on the History of Scottish Science." *History of Science* 12 (1974): 81-94.

———, and Arnold Thackray. *Gentlemen of Science: Early Years of the British Association for the Advancement of Science.* Oxford: Clarendon Press, 1981.

Morrison-Low, A. D., and J. R. R. Christie, eds. *"Martyr of Science": Sir David Brewster 1781-1868.* Edinburgh: Royal Scottish Museum, 1984.

Morse, E. W. "Natural Philosophy, Hypotheses, and Impiety: Sir David Brewster Confronts the Undulatory Theory of Light." PhD diss., University of California at Berkeley, 1972.

Morus, Iwan Rhys. *Bodies/Machines.* London: Berg, 2003.

———. "Correlation and Control: William Robert Grove and the Construction of a New Philosophy of Scientific Reform." *Studies in History and Philosophy of Science* 22 (1991): 589-621.

———. "The Electric Ariel: Telegraphy and Commercial Culture in Victorian England." *Victorian Studies* 39 (1996): 339-78.

———. "'The Nervous System of Britain': Space, Time, and the Electric Telegraph in the Victorian Age." *British Journal for the History of Science* 33 (2000): 455-75.

Moscucci, Ornella. *The Science of Woman: Gynaecology and Gender in England, 1800-1929.* Cambridge: Cambridge University Press, 1993.

Mullan, John. "Gendered Knowledge, Gendered Minds: Women and Newtonianism, 1690-1760." In *A Question of Identity: Women, Science, and Literature,* edited by Marina Benjamin. New Brunswick, NJ: Rutgers University Press.

Müller, Johannes. *Handbuch der Physiologie des Menschen für Vorlesungen.* 2 vols. Koblenz: J. Holscher, 1837-40.

———. *The Physiology of the Senses, Voice, and Muscular Motion, with the Mental Faculties.* Translated by William Baly. London: Taylor, Walton and Maberly, 1848.

———. *Zur vergleichenden Physiologie des Gesichtssinnes des Menschen und der Thiere.* Leipzig, 1826.

Neve, Michael. "Science in a Commercial City: Bristol 1820-60." In Inkster and Morrell, *Metropolis and Province.*

Newcomb, Simon. *Reminiscences of an Astronomer.* Boston: Houghton, Mifflin, 1903.

Newnham, William. *Essay on Superstition. . . .* London: J. Hatchard and Son, 1830.

Newton, Isaac. *The Correspondence of Isaac Newton.* Edited by H. W. Turnball. Cambridge: Cambridge University Press, 1961.

——. *Opticks; or, a Treatise of the Reflections, Refractions, Inflections, and Colours of Light.* 4th ed. 1730. Reprint, New York: Dover, 1952.

Nicholson, Cornelius. *Annals of Kendal.* 2nd ed. London, 1861.

Nicolai, Christoph Friedrich. "Memoir on the Appearance of Spectres or Phantoms Occasioned by Disease, with Psychological Remarks." *Journal of Natural Philosophy, Chemistry, and the Arts* 6 (Nov. 1803): 161–79.

Ogilvie, Marilyn Bailey. "Caroline Herschel's Contributions to Astronomy." *Annals of Science* 32 (1975): 149–61.

Olson, Richard. *Science Deified and Science Defied: The Historical Significance of Science in Western Culture.* 2 vols. Berkeley: University of California Press, 1990.

——. *Scottish Philosophy and British Physics, 1750–1880: A Study in the Foundations of the Victorian Scientific Style.* Princeton: Princeton University Press, 1975.

"On the Phantasms Produced by Disordered Sensation." *Journal of Natural Philosophy, Chemistry, and the Arts* 15 (1806): 288–96.

Orange, Derek. "The British Association for the Advancement of Science: The Provincial Background." *Science Studies* 1 (1971): 315–29.

——. *Philosophers and Provincials: The Yorkshire Philosophical Society from 1822 to 1844.* York: Yorkshire Philosophical Society, 1973.

——. "Rational Dissent and Provincial Science: William Turner and the Newcastle Literary and Philosophical Society." In Inkster and Morrell, *Metropolis and Province.*

Oreskes, Naomi. "Objectivity or Heroism? On the Invisibility of Women in Science." *Osiris* 11 (1996): 87–113.

Osler, Margaret. "Mixing Metaphors: Science and Religion or Natural Philosophy and Theology in Early Modern Europe." *History of Science* 35 (1997): 91–113.

Outram, Dorinda. *The Body and the French Revolution: Sex, Class, and Political Culture.* New Haven: Yale University Press, 1989.

Paley, William. *Natural Theology; or Evidences of the Existence and Attributes of the Deity, Collected from the Appearances of Nature.* 14th ed. London: J. Faulder, 1813.

——. *A View of Evidences of Christianity.* 3 vols. London: R. Faulder, 1794.

Palmer, George [Giros de Chantilly]. *Theory of Colors and Vision.* London: S. Leacroft, 1777.

Parish, Edmund. *Hallucinations and Illusions: A Study of the Fallacies of Perception.* London: Walter Scott; New York: Charles Scribner and Sons, 1898.

Parry, Caleb Hillier. *Collections from the Unpublished Medical Writings of the Late Caleb Hillier Parry.* Edited by Charles Henry Parry. London: Underwoods, 1825.

Parsons, Coleman O. *Witchcraft and Demonology in Scott's Fiction.* Edinburgh: Oliver and Boyd, 1964.

"Passages from the Diary of a Late Physician, ch. IV." *Blackwood's Edinburgh Magazine* 28 (1830): 770–93.

Patterson, Elizabeth C. *John Dalton and the Atomic Theory: The Biography of a Natural Philosopher*. Garden City, NY: Doubleday, 1970.

Pearsall, Ronald. *The Table-Rappers*. London: Joseph, 1972.

Pearson, Karl. *The Life, Letters and Labours of Francis Galton*. Cambridge: Cambridge University Press, 1914.

Péclet, Eugène. *Cours de physique*. Marseille: A. Ricard, 1823.

Penn, Roger. *Class, Power, and Technology: Skilled Workers in Britain and America*. Cambridge: Polity Press, 1990.

Peterfreund, Stuart. "Scientific Models in Optics: From Metaphor to Metonymy and Back." *Journal of the History of Ideas* 55 (1994): 59–73.

Pickstone, John V. "Bodies, Fields, and Factories: Technologies and Understandings in the Age of Revolutions." In *Technological Change: Methods and Themes in the History of Technology*, edited by Robert Fox. Amsterdam: Harwood Academic Publishers, 1996.

[Playfair, Lyon]. "The Relation of the Food of Man to His Muscular Power." *North British Review* 45 (1866): 321–43.

Pointon, Marcia. "Quakerism and Visual Culture 1650–1800." *Art History* 20 (1997): 397–431.

Pole, William. "Daltonism." *Contemporary Review* 37 (1880): 821–36.

——. "On Colour-Blindness." *Philosophical Transactions of the Royal Society of London* 149 (1859): 323–59.

——. "On the Present State of Knowledge and Opinion in regard to Colour-Blindness." *Transactions of the Royal Society of Edinburgh* 37 (1891–95): 441–79.

Poovey, Mary. *Making a Social Body: British Cultural Formation 1830–1864*. Chicago: University of Chicago Press, 1995.

Porter, Roy. "*Barely Touching:* A Social Perspective on Mind and Body." In Rousseau and Porter, *Languages of Psyche*.

——. "The Rage of Party: A Glorious Revolution in English Psychiatry?" *Medical History* 27 (1983): 35–50.

——, and Dorothy Porter. *In Sickness and in Health: The British Experience 1650–1850*. New York: Basil Blackwell, 1988.

Porter, Theodore M. "Objectivity as Standardization: The Rhetoric of Impersonality in Measurement, Statistics, and Cost-Benefit Analysis." *Annals of Scholarship* 9 (1992): 19–59.

——. "Precision and Trust: Early Victorian Insurance and the Politics of Calculation." In Wise, *Values of Precision*.

——. *The Rise of Statistical Thinking 1820–1900*. Princeton: Princeton University Press, 1986.

Powell, Baden. *An Historical View of the Progress of the Physical and Mathematical Sciences*. London: Longman, 1834.

Prescott, George B. *History, Theory, and Practice of the Electric Telegraph*. Boston: Ticknor and Fields, 1860.

Prest, John. *Liberty and Locality: Parliament, Permissive Legislation, and Ratepayers' Democracies in the Nineteenth Century*. Oxford: Clarendon, 1990.

Priestley, Joseph. *History and Present State of Discoveries Relating to Vision, Light, and Colours*. 2 vols. London: J. Johnson, 1772.

Purkyne, Jan Evangelista. *Beiträge zur Kenntniß des Sehens in Subjectiver Hinsicht*. Prague: Wildeubrunn, 1819.

Rabinbach, Anson. *The Human Motor: Energy, Fatigue, and the Origins of Modernity*. New York: Basic Books, 1990.

Raistrick, Arthur. *Quakers in Science and Industry.* . . . London: Bannisdale Press, 1950.

Reed, Edward S. *From Soul to Mind: The Emergence of Psychology from Erasmus Darwin to William James*. New Haven: Yale University Press, 1997.

Reed, Mark L. *Wordsworth: The Chronology of the Middle Years*. Cambridge: Harvard University Press, 1975.

Reid, Thomas. *Essays on the Active Powers of the Human Mind*. Edited by Baruch A. Brody. 1815. Reprint, Cambridge: MIT Press, 1969.

"Report of the Committee on Colour-Vision." *Proceedings of the Royal Society of London* 51 (1892): 281–396.

"Reports of the Sessions of the Société Médico-Psychologique." *Annales Médico-Psychologiques*, 3rd ser., 1 (1855): 541–44.

Ricardo, David. *Letters 1816–18*. Vol. 7 of *The Works and Correspondence of David Ricardo*. Edited by Piero Sraffa. Cambridge: Cambridge University Press, 1951.

Rider, Robin E. "Measure of Ideas, Rule of Language: Mathematics and Language in the 18th Century." In *The Quantifying Spirit in the Eighteenth Century*, edited by Tore Frängsmyr, J.L. Heilbron, and Robin E. Rider. Berkeley: University of California Press, 1990.

Riley, James C. *Sickness, Recovery, and Death: A History and Forecast of Ill Health*. Hampshire, UK: Macmillan, 1989.

Roberts, Lissa. "The Death of the Sensuous Chemist: The 'New' Chemistry and the Transformation of Sensuous Technology." *Studies in History and Philosophy of Science* 26 (1995): 503–29.

Robertson, J. Charles. "The Bacon-Facing Generation: Scottish Philosophy in the Early Nineteenth Century." *Journal of the History of Philosophy* 14 (1976): 37–49.

Rogers, E. C. *Philosophy of Mysterious Agents, Human and Mundane; or, the Dynamic Laws and Relations of Man*. Boston: John P. Jewett, 1853.

Roget, Peter Mark. "Explanation of an Optical Deception in the Appearance of the Spokes of a Wheel Seen through Vertical Apertures." *Philosophical Transactions of the Royal Society of London* 115 (1825): 131-40.

————. *An Introductory Lecture on Human and Comparative Physiology; Delivered at the New Medical School in Aldergate Street.* London, 1826.

Romano, Richard M. "The Economic Ideas of Charles Babbage." *History of Political Economy* 14 (1982): 385-405.

Roscoe, Henry E. *John Dalton and the Rise of Modern Chemistry.* London: Cassell, 1895.

Rothermel, Holly. "Images of the Sun: Warren De la Rue, George Biddell Airy and Celestial Photography." *British Journal for the History of Science* 26 (1993): 137-69.

Rousseau, G. S., and Roy Porter, eds. *The Languages of Psyche: Mind and Body in Enlightenment Thought.* Berkeley: University of California Press, 1987.

Rozier, François. "Observations sur quelque personnes qui ne peuvent distinguer les couleurs." *Journal de physique, de chimie, d'histoire naturelle et des arts* 13 (1779): 86-88.

Russell, Elbert. *The History of Quakerism.* New York: Macmillan, 1942.

Russett, Cynthia Eagle. *Sexual Science: The Victorian Construction of Womanhood.* Cambridge: Harvard University Press, 1989.

Sacks, Oliver. "Scotoma: Forgetting and Neglect in Science." In *Hidden Histories of Science,* edited by Robert B. Silvers. New York: New York Review of Books, 1995.

Sampson, Edward E. "Establishing Embodiment in Psychology." *Theory and Psychology* 6 (1996): 601-24.

Sandras, Claude-Marie-Stanislaus. *Traite pratique des maladies nerveuses.* Paris: Bailliere, 1851.

Sarbin, Theodore R., and Joseph B. Juhasz. "The Historical Background of the Concept of Hallucination." *Journal of the History of the Behavioral Sciences* 3 (1967): 339-58.

Schaaf, Larry. *Tracings of Light: Sir John Herschel and the Camera Lucida.* San Francisco: The Friends of Photography, 1989.

Schaffer, Simon. "Astronomers Mark Time: Discipline and the Personal Equation." *Science in Context* 2 (1988): 115-45.

————. "Experimenters' Techniques, Dyers' Hands, and the Electric Planetarium." *Isis* 88 (1997): 456-83.

————. "Machine Philosophy: Demonstration Devices in Georgian Mechanics." *Osiris* 9 (1994): 157-82.

————. "Physics Laboratories and the Victorian Country House." In *Making Space for Science,* edited by Crosbie Smith and Jonathan Agar. London: Macmillan, 1998.

————. "States of Mind: Enlightenment and Natural Philosophy." In Rousseau, *Languages of Psyche.*

Scherffer, Karl. *Dissertatione de Coloribus accidentalibus.* Vienna, 1761.

Schiebinger, Londa. *The Mind Has No Sex? Women in the Origins of Modern Science.* Cambridge: Harvard University Press, 1989.

Schiller, Francis. "The Migraine Tradition." *Bulletin of the History of Medicine* 49 (1975): 1–19.

Schimmelpenninck, Mary Anne Galton. *Life of Mary Anne Schimmelpenninck.* 2 vols. London: Longman, Brown, Green, Longmans, and Roberts, 1858.

Schofield, Robert. "Joseph Priestley, Eighteenth-Century British Neoplatonism, and Samuel Taylor Coleridge." In *Transformation and Tradition in the Sciences: Essays in Honor of I. Bernard Cohen,* edited by Everett Mendelsohn. Cambridge: Cambridge University Press, 1984.

———. *Mechanism and Materialism: British Natural Philosophy in an Age of Reason.* Princeton: Princeton University Press, 1970.

Scoresby, William. "An Inquiry into Some of the Circumstances and Principles Which Regulate the Production of Pictures on the Retina of the Human Eye." *Proceedings of the Royal Society of London* (1853): 380–83; (1854): 117–22.

Scott, J. "An Account of a Remarkable Imperfection of Sight; in a Letter from J. Scott to the Rev. Mr. Whisson, of Trinity College, Cambridge." *Philosophical Transactions of the Royal Society of London* 68 (1778): 611–14.

Scott, Walter. *The Journal of Sir Walter Scott.* New York: Harper and Bros., 1891.

———. *The Letters of Sir Walter Scott.* Edited by H. J. C. Grierson. London: Constable, 1936.

———. *Letters on Demonology and Witchcraft, Addressed to J. G. Lockhart, Esq.* Edited by Henry Morley. London: Routledge, 1884.

Scull, Andrew. "A Victorian Alienist: John Conolly, F.R.C.P., D.C.L. (1794–1866)." In Bynum, Porter, and Shepherd, *Anatomy of Madness.*

"Séances des 9 et 16 février." *Journal des Debats* (18 February 1846): 1.

Seashore, C. A. "Measurement of Illusions and Hallucinations in Normal Life." *Studies of the Yale Psychological Laboratory* 3 (1895): 1–67.

Secord, Anne. *Artisan Naturalists: Science as Popular Culture in Nineteenth-Century England.* Chicago: University of Chicago Press, forthcoming.

Seebeck, August. "Über den bei manchen Personen vorkommenden Mangel an Farbensinn." *Annalen der Physik und Chemie; von J. C. Poggendorff* 42 (1837): 177–233.

Seltzer, Mark. *Bodies and Machines.* New York: Routledge, 1992.

Sepper, Dennis L. *Goethe contra Newton: Polemics and the Project for a New Science of Color.* Cambridge: Cambridge University Press, 1988.

Shapin, Steven. "The Audience for Science in Eighteenth-Century Edinburgh." *History of Science* 12 (1974): 95–121.

———. "Brewster and the Edinburgh Career in Science." In Morrison-Low and Christie, *"Martyr of Science."*

———. "'Nibbling at the Teats of Science': Edinburgh and the Diffusion of Science in the 1830s." In Inkster and Morrell, *Metropolis and Province.*

———. *A Social History of Truth: Civility and Science in Seventeenth-Century England.* Chicago: University of Chicago Press, 1994.

Shapiro, Alan E. *Fits, Passions, and Paroxysms: Physics, Method, and Chemistry and Newton's Theories of Colored Bodies and Fits of Easy Reflection.* Cambridge: Cambridge University Press, 1993.

Sheepshanks, Richard. "Memoir of Edward Troughton." *Memoirs of the Royal Astronomical Society* 9 (1836): 283–90.

Sherman, Paul D. *Colour Vision in the Nineteenth Century: The Young-Helmholtz-Maxwell Theory.* Bristol, UK: Adam Hilger, 1981.

Showalter, Elaine. *The Female Malady: Women, Madness, and English Culture, 1830–1980.* New York: Pantheon, 1985.

———. *Hystories: Hysterical Epidemics and Modern Culture.* New York: Columbia University Press, 1997.

Sibum, Heinz Otto. "Working Experiments: Bodies, Machines, and Heat Values." In *The Physics of Empire,* edited by Richard Staley. Cambridge: Whipple Museum, 1997.

Sidgwick, Henry. "The Census of Hallucinations." *Proceedings of the Society for Psychical Research* 4 (1889–90): 7–25; 7 (1891–92): 429–35; 10 (1894): 25–252.

"Sir John Frederick William Herschel." *Daily News* (13 May 1871): 5–6.

Siraisi, Nancy G. *The Clock and the Mirror: Girolamo Cardano and Renaissance Medicine.* Princeton: Princeton University Press, 1997.

Smee, Alfred. *Vision in Health and Disease; the Value of Glasses for Its Restoration, and the Mischief Caused by Their Abuse.* London: Horne, Thornthwaite, and Wood, 1847.

Smiles, Samuel. *Lives of the Engineers.* 1862. Reprint, New York: Augustus M. Kelley, 1968.

Smith, Crosbie. "Mechanical Philosophy and the Emergence of Physics in Britain, 1800–1850." *Annals of Science* 33 (1976): 3–29.

———, and M. Norton Wise. *Energy and Empire: A Biographical Study of Lord Kelvin.* Cambridge: Cambridge University Press, 1989.

Smith, Robert W. "A National Observatory Transformed: Greenwich in the Nineteenth Century." *Journal for the History of Astronomy* 22 (1991): 5–20.

Smith, Roger. "The Background of Physiological Psychology in Natural Philosophy." *History of Science* 11 (1973): 75–123.

———. *Norton History of the Human Sciences.* New York: W. W. Norton, 1997.

Smith, William. *A Dissertation upon the Nerves*. London: Printed for the author, 1768.

Smyth, W. H. *A Cycle of Celestial Objects for the Use of Naval, Military, and Private Astronomers*. 2 vols. London: J. W. Parker, 1844.

Somerville, Mary. *Personal Recollections, from Early Life to Old Age, of Mary Somerville; with Selections from her Correspondence*. Edited by Martha Somerville. Boston: Roberts Brothers, 1874.

Sommer, Karl (?). "Über Chromatopseudopsie, oder den, manchen Menschen eigenen Mangel des Farbenunterscheidungsvermögens." *Journal der Chirurgie und Augen-Heilkunde* 5 (1823): 19-43.

Sorrenson, Richard J. "Towards a History of the Royal Society in the 18th Century." *Notes and Records of the Royal Society of London* 50 (1996): 29-46.

Southey, Robert. "History of Dissenters." *Quarterly Review* 10 (1813): 90-139.

———. *Sir Thomas More; or Colloquies on the Progress and Prospects of Society*. London: J. Murray, 1829.

Spencer, Herbert. *An Autobiography*. Vol. 20-21 of *The Works of Herbert Spencer*. 1904. Reprint, Osnabrück: Otto Zeller, 1966.

———. *The Principles of Psychology*. 2 vols. 3rd ed. New York: D. Appleton and Company, 1905.

Sprat, Thomas. *The History of the Royal-Society of London, for the Improving of Natural Knowledge*. London, 1667.

Stafford, Barbara Maria. *Body Criticism: Imaging the Unseen in Enlightenment Art and Medicine*. Cambridge: MIT Press, 1991.

———. *Voyage into Substance: Art, Science, Nature and the Illustrated Travel Account*. Cambridge: MIT Press, 1984.

Stafford, Robert A. "Geological Surveys, Mineral Discoveries, and British Expansion, 1835-71." *Journal of Imperial and Commonwealth History* 12, no. 3 (1984): 5-32.

Standage, Tom. *The Victorian Internet*. London: Phoenix, 1998.

Staum, Martin S. *Cabanis: Enlightenment and Medical Philosophy in the French Revolution*. Princeton: Princeton University Press, 1980.

Stephen, Leslie, ed. *Dictionary of National Biography*. London: Smith, Elder, and Co., 1885-1901.

Stewart, Balfour, and P. G. Tait. *The Unseen Universe; or Physical Speculations on a Future State*. London: Macmillan, Constable, 1875.

Stewart, Dugald. *Elements of the Philosophy of the Human Mind*. Vol. 2 of *The Collected Works of Dugald Stewart*. Edited by William Hamilton. Edinburgh: T. & T. Clark, 1877.

———. "Some Account of a Boy Born Blind and Deaf, Collected from Authentic Sources of Information; with a Few Remarks and Comments." *Transactions of the Royal Society of Edinburgh* 7 (1814): 1-78.

Strickland, Stuart. "The Ideology of Self-Knowledge and the Practice of Self-Experimentation." *Eighteenth-Century Studies* 31 (1998): 453–71.

Struve, Otto W. *Wilhelm Struve: Zur Erinnerung an der Vater den Geschwistern dargebracht von Otto Struve.* Karlsruhe: Druck der Braun'schen Hofbuchdruckerie, 1895.

Sussman, Herbert L. *Victorians and the Machine: The Literary Response to Technology.* Cambridge: Harvard University Press, 1968.

Suzuki, Akihito. "Dualism and the Transformation of Psychiatric Language in the Seventeenth and Eighteenth Centuries." *History of Science* 33 (1995): 417–47.

Swift, Jonathan. *Gulliver's Travels.* Vol. 8 of *The Prose Works of Jonathan Swift.* Edited by Temple Scott. London: George Bell and Sons, 1899.

Swijtink, Zeno G. "The Objectification of Observation." In Krüger, *The Probabilistic Revolution.*

Symondson, Anthony, ed. *The Victorian Crisis of Faith: Six Lectures.* London: Society for Promoting Christian Knowledge, 1970.

Szokalski, Wiktor Felix. "Essai sur les sensations des couleurs dans l'état physiologique et pathologique de l'oeil." *Annales d'Occulistique* 2 (Oct. 1839): 11–21 (Nov. 1839): 37–50 (Jan. 1840): 77–92 (March 1840): 165–77; 3 (April 1840): 1–20 (May 1840): 49–64 (June 1840): 97–122.

Tait, P. G. *Lectures on Some Recent Advances in Physical Science; with a Special Lecture on Force.* 2nd ed. London: Macmillan, 1876.

Tannoch-Bland, Jennifer. "Dugald Stewart on Intellectual Character." *British Journal for the History of Science* 30 (1997): 307–20.

Tattersall, J. J. "Nicholas Saunderson, The Blind Lucasian Professor." *Historia Mathematica* 19 (1992): 356–70.

Taylor, John. *Account of the Mechanism of the Eye; with an Endeavour to Ascertain the True Place of a Cataract.* Norwich: Henry Crossgrove, 1727.

Taylor, Joseph. *Apparitions; or, The Mystery of Ghosts, Hobgoblins, and Haunted Houses, Developed.* 2nd ed. London: Lackington, Allen, 1815.

Thackray, Arnold. *Atoms and Powers: An Essay on Newtonian Matter Theory and the Development of Chemistry.* Cambridge: Harvard University Press, 1970.

———. *John Dalton: Critical Assessments of His Life and Science.* Cambridge: Harvard University Press, 1972.

———. "Natural Knowledge in Cultural Context: The Manchester Model." *American Historical Review* 79 (1974): 672–709.

Thiebault, Dieudonne. *Original Anecdotes of Frederic the Second, King of Prussia. . . .* London: J. Johnson, 1805.

Thomson, William. "The Bangor Laboratories." In *William Thomson, Popular Lectures and Addresses.* 2nd ed. London, 1891.

Thurston, Robert H. *A History of the Growth of the Steam-Engine*. New York: D. Appleton, 1878.

Tresch, John. "Mechanical Romanticism: Engineers of the Artificial Paradise." PhD thesis, University of Cambridge, 2001.

Trotter, Thomas. *A View of the Nervous Temperament*. London: Longman, Hurst, Rees, and Orme, 1807.

Troughton, Edward. "An Account of a Method of Dividing Astronomical and Other Instruments, by Ocular Inspection." *Philosophical Transactions of the Royal Society of London* 99 (1809): 105-45.

Tuana, Nancy. *The Less Noble Sex: Scientific, Religious, and Philosophical Conceptions of Woman's Nature*. Bloomington: Indiana University Press, 1993.

Tubervile, Dawbeney. "Two Letters from the Great, and Experienced Oculist Dr. Turberville [*sic*] of Salisbury, . . . containing Several Remarkable Cases of Physick, Relating Chiefly to the Eyes." *Philosophical Transactions of the Royal Society of London* 14 (1684): 736-38.

Tuke, D. H. "Hallucinations, and the Subjective Sensations of the Sane." *Brain* 11 (1888-89): 441-46.

Turner, Frank M. *Between Science and Religion: The Reaction to Scientific Naturalism in Late-Victorian England*. New Haven: Yale University Press, 1974.

Turner, R. Steven. *In the Eye's Mind: Vision and the Helmholtz-Hering Controversy*. Princeton: Princeton University Press, 1994.

———. "Paradigms and Productivity: The Case of Physiological Optics, 1840-94." *Social Studies of Science* 17 (1987): 35-68.

Tuttle, Julianne. "Humphry Davy: A Case Study in Science and Romanticism." PhD diss., Indiana University, 2000.

Tuveson, Ernest L. *The Imagination as a Means of Grace: Locke and the Aesthetics of Romanticism*. Berkeley: University of California Press, 1960.

Twining, William. "On Single Vision and the Union of the Optic Nerves." *Edinburgh Journal of Science* 9 (1828): 143-53.

Tyndall, John. "Coloured Lamps on Railways." *Athenaeum* no. 1318 (29 January 1853): 136-37.

———. *Faraday as a Discoverer*. New York: D. Appleton and Co., 1880.

———. *Heat Considered as a Mode of Motion: Being a Course of Twelve Lectures Delivered at The Royal Institution of Great Britain in the Season of 1862*. London, 1863.

———. *Mountaineering in 1861: A Vacation Tour*. London, 1862.

———. "On a Peculiar Case of Colour Blindness." *Philosophical Magazine* 11 (1856): 329-33.

Tyrrell, Frederick. *Practical Work on the Diseases of the Eye, and Their Treatment, Medically, Topically, and by Operation.* London: J. Churchill, 1840.

Ure, Andrew. *The Philosophy of Manufactures; or, an Exposition of the Scientific, Moral, and Commercial Economy of the Factory System of Great Britain.* 3rd ed. London: H. G. Bohm, 1861.

Van Helmont, Jean Baptiste. *Ortus medicinae: id est, initia physiciae inaudita; progressus medicinae novus, in morborum ultionem, ad vitam longam.* Amsterdam: Apud Ludovicum Elsevirium, 1648.

Veitch, John. "Memoir of Dugald Stewart." In vol. 10 of Stewart, *Collected Works.*

Voller, Jack G. *The Supernatural Sublime: The Metaphysics of Terror in Anglo-American Romanticism.* Dekalb: Northern Illinois University Press, 1994.

Vrettos, Athena. *Somatic Fictions: Imagining Illness in Victorian Culture.* Stanford: Stanford University Press, 1995.

Wach, Howard M. "Culture and the Middle Classes: Popular Knowledge in Industrial Manchester." *Journal of British Studies* 27 (1988): 375-404.

Wade, Nicholas. *Brewster and Wheatstone on Vision.* London: Academic Press, 1983.

Wahrman, Dror. *Imagining the Middle Class: The Political Representation of Class in Britain, c. 1780-1840.* Cambridge: Cambridge University Press, 1995.

———. "National Society, Communal Culture: An Argument about the Recent Historiography of Eighteenth-Century Britain." *Social History* 17 (1992): 43-72.

Walker, Ezekiel. "Description of Mr. Ez. Walker's New Optical Machine Called the Phantasmascope." *Philosophical Magazine* 27 (1807): 97-98.

Wardrop, James. *Essays on the Morbid Anatomy of the Human Eye.* London: Archibald Constable, 1818.

Warner, John Harley. "The Idea of Science in English Medicine: The 'Decline of Science' and the Rhetoric of Reform, 1815-45." In French and Wear, *British Medicine.*

Wartmann, Elie. "Memoir on Daltonism." In vol. 4 of *Scientific Memoirs Selected from the Transactions of Foreign Academies of Science and Learned Societies, and from Foreign Journals.* Edited by Richard Taylor. London: Richard and John E. Taylor, 1846.

Warwick, Andrew. "Exercising the Student Body: Mathematics and Athleticism in Victorian Cambridge." In Lawrence and Shapin, *Science Incarnate.*

Watkins, John, and Frederic Shoberl. *A Biographical Dictionary of the Living Authors of Great Britain and Ireland.* 1816. Reprint, Detroit: Gale Research Co., 1966.

Werner, Abraham Gottlob. *Werner's Nomenclature of Colours.* Translated by Patrick Syme. Edinburgh: W. Blackwood, 1814.

Wheatstone, Charles. "Contributions to the Physiology of Vision, no. I." *Journal of the Royal Institution* 1 (1830): 101-17.

Whewell, William. *Astronomy and General Physics Considered with Reference to Natural Theology*. London: W. Pickering, 1839.

———. *History of the Inductive Sciences, from the Earliest to the Present Time*. 3rd ed. London, 1857.

———. *The Philosophy of the Inductive Sciences*. 2nd ed. London: J. W. Parker, 1847.

Whiston, William. *Account of the Exact Time When Miraculous Gifts Ceas'd in the Church*. London: printed for the author, 1749.

Whytt, Robert. *Observations on the Nature, Causes and Cure of Those Diseases Which Are Commonly Called Nervous, Hyperchondriac or Hysteric. . . .* Edinburgh, 1764.

Wilkinson, J. Gardiner. *On Colour, and on the Necessity for a General Diffusion of Taste among the Classes*. London, 1858.

Wilson, George. "Dalton." *Fraser's Magazine* 50 (Nov. 1854): 554–72.

———. "On Colour-Blindness; in Connexion with the Employment of Coloured Signals on Railways." *Athenaeum* no. 1327 (2 April 1853): 418.

———. "On the Prevalence of Chromato-pseudopsis or Colour-Blindness; Its Evils, and the Means of Diminishing Its Frequency." *Edinburgh Monthly Journal of Medical Science* 17 (1853): 377–96, 491–507; 18 (1854): 37–44, 309–25, 411–16; 19 (1854): 1–10, 97–107, 226–40, 393–403, 490–504.

———. *Religio Chemici*. London: Macmillan, 1862.

———. *Researches on Colour-Blindness; with a Supplement on the Danger Attending the Present System of Railway and Marine Coloured Signals*. Edinburgh: Sutherland and Knox, 1855.

Wilson, James. *Biography of the Blind; including the Lives of All Who Have Distinguished Themselves as Poets, Philosophers, Artists, &c.* 2nd ed. Birmington: J. W. Showell, 1833.

Wilson, Jessie A. *Memoir of George Wilson, M.D., F.R.S.E., Regius Professor Technology in the University of Edinburgh and Director of the Industrial Museum of Scotland*. Edinburgh: Edmonton and Douglas, 1860.

Winslow, Forbes. *Anatomy of Suicide*. London: H. Renshaw, 1840.

Winter, Alison. "A Calculus of Suffering: Ada Lovelace and the Bodily Constraints on Women's Knowledge in Early Victorian England." In Lawrence and Shapin, *Science Incarnate*.

———. *Mesmerized: Powers of Mind in Victorian Britain*. Chicago: University of Chicago Press, 1998.

Wise, M. Norton. "Mediating Machines." *Science in Context* 2 (1988): 77–113.

———. "Precision: Agent of Unity and Product of Agreement, Part 2—The Age of Steam and Telegraphy." In Wise, *Values of Precision*.

———, ed. *The Values of Precision*. Princeton: Princeton University Press, 1995.

——, and Crosbie Smith. "Measurement, Work, and Industry in Lord Kelvin's Britain." *Historical Studies in the Physical and Biological Sciences* 17 (1986): 147-73.

——, with Crosbie Smith. "Work and Waste: Political Economy and Natural Philosophy in Nineteenth Century Britain." *History of Science* 27 (1989): 263-301, 391-449; 28 (1990): 221-61.

Wollaston, William Hyde. "On an Improvement in the Form of Spectacle Glasses." *Philosophical Magazine* 17 (1803): 327-29.

——. "On the Semi-decussation of the Optic Nerves." *Philosophical Transactions of the Royal Society of London* 114 (1824): 222-45.

Wollstonecraft, Mary. *A Vindication of the Rights of Women*. Edited by Carol H. Poston. 1792. Reprint, New York: 1975.

Woods, Robert. "Physician, Heal Thyself: The Health and Mortality of Victorian Doctors." *Social History of Medicine* 9 (1996): 1-30.

Wordsworth, William. "The Excursion." In vol. 5 of *The Poetical Works of William Wordsworth*, edited by E. de Selincourt and Helen Darbishire. Oxford: Clarendon, 1959.

Wright, W. D. "The Unsolved Problem of 'Daltonism.'" In *John Dalton and the Progress of Science*, edited by D. S. L. Cardwell. Manchester: Manchester University Press; New York: Barnes and Noble, 1968.

Yeo, Richard. *Defining Science: William Whewell, Natural Knowledge, and Public Debate in Early Victorian Britain*. Cambridge: Cambridge University Press, 1993.

——. "Genius, Method, and Morality: Images of Newton in Britain, 1760-1860." *Science in Context* 2 (1988): 257-84.

——. "An Idol of the Marketplace: Baconianism in Nineteenth-Century Britain." *History of Science* 23 (1985): 251-98.

——. "Reviewing Herschel's *Discourse*." *Studies in History and Philosophy of Science* 20 (1989): 541-52.

——. "Scientific Method and the Image of Science 1831-1891." In MacLeod and Collins, *Parliament of Science*.

——. "Scientific Method and the Rhetoric of Science in Britain, 1830-1917." In *The Politics and Rhetoric of Scientific Method: Historical Studies*, edited by John A. Schuster and Richard R. Yeo. Dordrecht: D. Reidel, 1986.

——. "William Whewell, Natural Theology and the Philosophy of Science in Mid Nineteenth Century Britain." *Annals of Science* 36 (1979): 493-516.

Young, Robert M. *Mind, Brain, and Adaptation in the Nineteenth Century: Cerebral Localization and Its Biological Context from Gall to Ferrier*. Oxford: Clarendon Press, 1970.

Young, Thomas. *A Course of Lectures on Natural Philosophy and the Mechanical Arts*. 2 vols. London: J. Johnson, 1807.

——. "On the Mechanism of the Eye." *Philosophical Transactions of the Royal Society of London* 91 (1801): 23-88.

Index

Abercrombie, John, 44, 163
accidental colors. *See* hemiopsy
achromatism, 64–65
Adams, John Couch, 113
Admiralty, 128, 191
Aepinus, Franz, 124
afterimages, 10, 62–63, 105, 124, 138
Agassi, Joseph, 133
Airy, George, 82, 143–45; astigmatism of, 128, 131; hemiopsy of, 103, 105, 115, 128–34, 138, 140; observational practices of, 22, 35–36, 123, 127–28; on women in the sciences, 45
Airy, Hubert, 16–17, 29, 220n118, 220n123; on hemiopsy, 103, 105, 115, 138–41
Airy, Richarda, 133, 138
Airy, Wilfrid, 126–27
Albert, Prince, 57, 99
Alison, William Pulteney, 69
Ampère, André-Marie, 142
Analytical Society, 48, 152
animals: compared to humans, 34, 59
aphasia, 129, 140
apparitions. *See* ghosts; hallucination
Arago, François, 28–29, 145; hemiopsy of, 10, 105, 140
Arnold, Thomas, 160
associationism, 15, 20
astigmatism, 128, 131
astrology, 165
astronomy: importance of color in, 62–63; observational practices in, 19, 22, 28–29, 35–36, 47–48, 127–28

atheism, 149, 151–52, 155, 187
attention, 19, 22–23, 43, 179, 192
Augustine, 6

Babbage, Charles, 48, 109, 120–21; and John Dalton, 76–77; hallucinations of, 178–79; nervous breakdown of, 4, 113; on color blindness, 57, 210n80; on division of labor, 33, 35, 143; on governors, 110; on overwork, 114; on state support for the sciences, 41, 77; philosophy of science of, 22, 37, 143
Babington, George, 117
Bacon, Francis, 189; on longevity, 25; on women in the sciences, 45
Baconianism, 32, 195
Bailly, John Sylvain, 10, 28–29
Baily, Francis, 26, 40
balance, 43, 102–6, 115, 119, 142
Balfour, J. H., 112–13
Balfour, Stewart, 186
Banks, Joseph, 40
Barère de Vieuzac, Bertrand, 190
Beaufort, Francis, 40
Beddoes, Thomas, 177
Bell, Charles, 15, 27
Bennett, Risdon, 184
Berrios, German, 147
Bessel, Friedrich, 36
biblical criticism, 50, 193
Bickerton, Thomas Herbert, 88
Biot, Jean-Baptiste, 27
Birmingham Philosophical Society, 182

blindness, 140, 207n4; and scientific methodology, 55–56; as metaphor for provincialism, 69–70; caused by scientific work, 10, 29, 63

blindness, color. *See* color blindness

blind spot (*punctum caecum*), 23, 116, 131, 141

Board of Longitude, 14

body, human, 172, 174; compared to machine, 9, 15–16, 24, 106–12, 123, 144; historiography of, 23–24; importance in natural philosophy, 23–29; of scholars, 25, 112–14

Boerhaave, Herman, 10

Bonnet, Charles, 10, 174

Booth, J. Gore, 184

Bordo, Susan, 23

Boring, Edwin, 32

Bostock, John, 175

Boulton, Matthew, 110

Boyle, Robert, 68, 153

Breen, Hugh, 127

Brewster, David, 55, 67, 96, 109, 128, 132, 143–45, 188; hallucinations of, 169; hemiopsy of, 103, 105, 115, 122–26, 128–29, 131, 140; on color blindness, 50, 74, 87, 91; on hallucination, 16, 165, 167–70, 173, 181; on John Dalton, 72, 76; on *muscae volitantes*, 125–26; on Newton's sanity, 27–28; philosophy of science of, 12, 37, 143, 208n38; problems with vision, 63; religious views of, 14, 153, 155, 170, 200n39

Brierre de Boismont, Alexandre, 170–72, 174–75

Bristol, 41–42

British and Foreign Bible Society, 152

British Association for the Advancement of Science, 14, 29, 37, 40–41, 71, 76–77, 112–13, 137, 151

Brocken specter, 169–70

Brodie, Benjamin, 117, 158, 179

Brougham, Henry, 155

Broussais, François-Joseph-Victor, 103

Brown, Samuel, 18, 196

Brown, Thomas, 20, 158, 196

Browne, Thomas, 159

Bryce, William, 212n116

Bryson, J. M., 92, 95

Buffon, Georges-Louis Leclerc, Comte de, 63, 105

Burke, Edmund, 151

Burney, Frances, 58

Bushmen. *See* San

Byron, George Gordon, 6th Baron, 173

Cabanis, Pierre-Jean-Georges, 103, 174, 195

Cambridge University, 14, 42, 45, 48, 104, 113, 127, 152, 156, 178

camera lucida, 125, 220n118

camera obscura, 21, 67

Campbell, Lewis, 90

Cannon, Susan Faye, 9

Cape Colony, 34, 137

Cardano, Girolamo, 173

Carlyle, Thomas, 20, 108, 146

Carpenter, Elias, 160

Carpenter, William, 22, 111, 154

case histories, 79–81, 84, 86–87, 89, 100, 196

Castle, Terry, 157

cataract, 110

Catholicism, 150, 164

Chadwick, Edwin, 40

Challis, James, 113

Chalmers, Thomas, 172

Chappe, Claude, 191, 227n3

Chappe, Ignace, 192

chemistry: emergence as a discipline, 194; importance of color in, 62; observational practices in, 18

Chevreul, M. E., 68, 97

Cheyne, George, 6–7, 10, 25–26

Christianity, 6, 60–61; and hallucination, 146–88; and reason, 9, 132–33, 147, 150–54, 163–64, 171–72; Broad Church Anglicanism, 9, 14, 147, 149–55; Dissenting churches, 9, 39, 41, 72, 78, 147, 149–51; evangelicalism, 9, 75, 148, 152–53, 155–56, 187, 200n39; impact on natural philosophy, 34, 132–33, 186–87, 195. *See also*

atheism; biblical criticism; Catholicism; Church of England; deism; miracles; natural theology; nonconformity; Quakers; toleration, religious; Unitarians; voluntarism
Church of England, 9, 149
Clerke, Agnes Mary, 113
Clifford, William Kingdon, 111
Coleridge, Samuel Taylor, 27, 43, 146, 177
Coleridge, Sara, 109
color: importance of in natural philosophy, 62–65; physical explanations of, 63–65
color blindness, 55–100, 136, 186, 196; and problem of communication, 50; and railway safety, 86–88, 100; and taste, 50, 58–59, 69, 100; as metaphor for provincialism, 8, 57–62, 190; causes, 63, 66–67, 72, 82, 91, 96–98; compared to hallucination, 181–82, 184; frequency of, 86, 88; tied to Quakers, 56, 69, 72, 74–79
color vision: experimental tests on, 67–68, 79–86, 91–93; physiology of, 62–63
Colquhoun, Hugh, 58
Combe, Andrew, 101
Combe, George, 66, 143
Committee on Colour Vision (Royal Society), 87–88
Common Sense school, 14–15, 20, 158, 163, 196
communication. See language, clear
Condorcet, Marie Jean Antoine Nicolas Caritat, marquis de, 179
Conolly, John, 174
Cooke, William Fothergill, 190, 192
Cooper, White, 57
Crichton, Alexander, 160
Crosland, Newton, 160
Cullen, William, 15, 32–33, 196
Cunningham, Andrew, 194

Dalton, Deborah, 75
Dalton, John, 43, 55–56; as a Quaker, 75–79; as heroic scientific figure, 59, 61–62, 71, 76–79, 82, 90, 99; autopsy of, 72–73;

color blindness of, 8, 50, 59, 66, 70–72, 81, 84–85, 97–99, 209n63, 210n80
Dalton, Jonathan, 66, 70
Dalton, Joseph, 74
Darwin, Charles, 60–61; illness of, 11, 44; on emotion, 43–44
Darwin, Erasmus, 105, 207n5, 215n12
Darwin, Horace, 183
Darwin, Robert Waring; on color vision, 63, 215n12; on hemiopsy, 105–6, 117, 138
Darwinism, 195
Daston, Lorraine, 36, 80, 189
Davy, Humphry, vi, 30, 78, 120, 153; illness of, 4; nitrous oxide experiments of, 4, 175, 177; on provincialism, 40, 78
deduction, 19
Defoe, Daniel, 159–60
deism, 148–49, 152, 153, 164, 187, 195. See also voluntarism
de La Hire, Philippe, 68, 126
demonology, 165–66, 172
de Morgan, Augustus, 138, 156, 214n146
de Morgan, Sophia, 138
Dendy, Walter Cooper, 150, 164–65, 174
De Quincey, Thomas, 175, 177
Descartes, René, 153, 173
determinism, 104, 108. See also will
Dickens, Charles, 63
Dickerson, Vanessa, 146–47
difference engine, 33, 48
division of labor, 5, 12, 31, 33–34, 39, 101–2, 127–28, 141–45, 192–93; gendering of in the sciences, 44–48
Dollond, George, 81
Dollond, John, 64
Dombey and Son, 63
dualism. See mind-body dualism
Dufour, Guillaume-Henri, 129–31, 140
Dulong, Pierre-Louis, 145
dyspepsia, 124
Dyster, Frederic, 50, 74

Edelcrantz, Abraham, 191, 227n9
Edgeworth, Maria, 137, 227n9
Edgeworth, Richard Lovell, 227n9

Edinburgh, 42
Edinburgh Review, 42
Edinburgh, University of, 14, 42, 76, 86, 91, 112
efficiency, 9, 14, 32, 43, 101–3, 127, 193
electricity, 154; compared to nervous impulses, 16. *See also* telegraph, electric
Elias, Norbert, 23
Elliot, John, 15
empire, British, 193; impact on scientific methodology, 34, 59–60
empiricism, 13, 32, 49–51
Encke, Johann Franz, 36
Enlightenment, 151; compared to industrial age, 13, 31, 43, 102–6, 142, 145
error. *See* vision, subjectivity of
Esquirol, Etienne, 160–61, 171
ether, 64, 155–56
exercise, 126–27, 138; importance for health, 27; preventing nervous disorder, 112–14, 119, 131–32
experiment, 45, 51; as defining element in natural philosophy, 11, 196–97; compared to phenomenology, 32; physiological effects of, 28–29, 105, 123–24, 131, 144; problem of replication, 49, 97; used to test vision, 67–68, 79–86, 91–93, 168
expertise, 31
eye, 168; anatomy and physiology of, 65–67, 117, 118, 122–23, 126; imperfection of, 19, 27, 172
Eyre controversy, 59–60

faculty psychology. *See* associationism, attention, imagination, phrenology, reason, will
fairies, 165
Faraday, Michael, 121, 156; nervous breakdown of, 113; on managing natural philosopher's body, 44; philosophy of science of, 44
fatigue. *See* overwork
Fechner, Gustav, 80; problems with vision, 10, 63, 124

Ferriar, John, 161–62, 187
Ficinus, Marsilio, 173
Forbes, James, 91
forces, imponderable, 15–16, 111, 144, 154–55. *See also* electricity; gravitation; magnetism; optics
Foster, Joseph, 84–85
Fothergill, John, 140
Foucault, Michel, 23
Fox, George, 74
France: scientific culture compared to Britain's, 11, 29, 154
Franklin, Benjamin, 104, 179
French Revolution, 32, 149, 151, 154, 164, 187, 190–91; impact on natural philosophy, 11, 31
Fresnel, Augustin, 62

Galton, Francis: hallucinations of, 179, 181; nervous breakdown of, 113–14; on hallucination, 181–86
Galvani, Luigi, 16
Gamble, John, 191
Garnett, William, 90
Gaskins, Thomas, 207n4
Gassendi, Pierre, 173
gender. *See* masculinity; women
genius, 120–21; and insanity, 28, 179; and provincialism, 71, 89–90
Gentlemen's Diary, 43
Geological and Polytechnic Society (Yorkshire), 40
Geological Surveys, 40
George III, King, 58
ghosts, 9, 146–47, 153, 155, 157, 165–69, 178, 188. *See also* hallucination
Gladstone, J. H., 123–24
Gladstone, William, 59
Glaisher, James, 127
Glanvill, Joseph, 151
Gleditsch, J. G., 166
Goethe, Johann Wolfgang von, 174; on color vision, 16, 63
Gooding, David, 49
Gordon, James, 156

Gordon, Margaret Maria, 169
Gough, John, 71, 207n4; as heroic scientific figure, 55-56, 61-62
governors, 8-9, 106, 110-11, 141, 227n9
gravitation, 152, 154-55
Great Exhibition, 57, 189
Great Northern Railway Company, 87
Greenwich Observatory. *See* Royal Observatory, Greenwich
Gregory, Samuel, 163
Grove, William Robert, 111
Gruthuisen, Franz von Paula, 174
Guerrini, Anita, 6
Gulliver's Travels, 48-49
Gurney, Edmund, 186

Haley, Bruce, 23-24
Haller, Albrecht von, 15
hallucination, 9, 106, 146-88, 190; and emergence of psychology, 170-72; and imagination, 157-59; and nervous system, 162-63; compared to color blindness, 181-82, 184; etymology of, 159; medicalization of, 147-48, 159-60; natural philosophers' experiences of, 146, 169, 173-85; popular literature on, 159-70. *See also* ghosts
Hamilton, Grace, 36
Hamilton, Robert, 142
Hamilton, William, 21, 32-33
Hamilton, William Rowan, 35-36
Hansteen, Christopher, 132
Hartley, David, 15, 103, 171
Harvey, George, 80-81
Hatfield, Gary, 228n30
Hawking, Stephen, 25
Hawkins, Anne Hunsaker, 6
Hay, David Ramsay, 68, 91
Heineken, N. S., 122
Helmholtz, Hermann von, 96, 98-99; migraines of, 10; philosophy of science of, 22-23, 50-51, 111
hemiopsy, 8, 10, 101-45, 168, 173, 190; defined, 101, 220n123
hemiplegia, 133-34, 138, 140, 219n102

Henry, John, 151
Henry, Joseph, 10
Henry, William, 59, 120-21; and John Dalton, 77; on managing natural philosopher's body, 28; suicide of, 29, 112
Henslow, George, 179-80
heredity, 184-85
Hering, Ewald, 63
Herschel, Caroline, 47-48, 132
Herschel, Isabella, 136, 182-83
Herschel, John F. W., 40, 98-99, 113, 123, 128, 144, 156, 172; hallucinations of, 178; hemiopsy of, 103, 105, 115, 132-38, 140; on color blindness, 67, 81-86, 91, 93-94, 96-97, 100, 136, 214n146; on division of labor, 39, 143; on hallucination, 177-78; on managing natural philosopher's body, 26; on women in the sciences, 45-47, 168; philosophy of science of, 12, 35, 39, 65, 132-33, 143, 159
Herschel, John (J. F. W. Herschel's son), 183
Herschel, Margaret, 137, 177
Herschel, William, 47, 83, 132, 137, 178; nervous breakdown of, 4
Hibbert, Samuel, 162-63, 169, 175
historiography: of science, 3, 5, 7, 12, 24, 36, 189-90, 194-97; of the body, 23-24; of nineteenth-century Britain, 3, 189-90
Hogg, Jabez, 66, 207n5
Holland, Henry, 16, 37-38, 77-78
Homer, 59
Hooke, Robert, 227n3
Huddart, Joseph, 69, 74
Hume, David, 20, 103-4, 150-51, 153, 158-59, 222n15; nervous breakdown of, 26
Hutchison, Keith, 64
Huxley, Thomas Henry: on bodies as machines, 111-12
hypochondriasis, 4, 174. *See also* hysteria; nervous disorder
hysteria, 3, 45, 150, 171. *See also* hypochondriasis; nervous disorder

idiosyncrasy. *See* objectivity; provincialism
illusion, optical, 171, 187. *See also* Brocken
    specter
imagination, 20, 44, 157–59, 161, 163, 165,
    167–68, 170, 174–75, 179, 183, 185, 187
inductive hierarchy, 34–37
industrialization: effects on sciences, 5,
    10–11, 34–35, 39, 101–3, 106–12, 141–45
insanity. *See* madness
instruments, scientific, 11, 51, 64, 81, 92–
    93, 94–95, 104, 196; self-registering, 18
International Congress of Psychology, 186
Ivory, James, 78

Jago, James, 172
Jevons, W. S., 183
Johnson, Samuel, 173–74
Jones, Richard, 35
Joule, James, 153
Jurin, James, 27

kaleidoscope, 167
Kater, Henry, 122
Kelland, Philip, 87
Keller, Evelyn Fox, 23
Kelvin, Lord. *See* Thomson, William
King, Joshua, 207n4
Kitchiner, William, 122

Lacaille, Nicolas-Louis de, 28–29
Ladd-Franklin, Christine, 98–99
Ladies' Diary, 43
Lagrange, Joseph-Louis, 152
la Mettrie, Julien, 154
Lang, Andrew, 146
language, clear, 44, 48–51, 72, 74–75, 85–
    86, 92, 94–95, 97, 191–93
Laplace, Pierre-Simon, 46, 104, 152
latitudinarianism. *See* Christianity: Broad
    Church Anglicanism
laudanum, 136, 219n108. *See also* opium
Lavater, Ludwig, 159
Lefroy, Henry, 183
Letters on Demonology and Witchcraft, 165–
    68

Letters on Natural Magic, 16, 123, 167–70
Lewes, George Henry, 111
Lister, J. J., 84
Literary and Philosophical Societies, 39–
    40; Leeds, 137–38; Manchester, 39, 41,
    71, 187
Lloyd, Genevieve, 23
Lloyd, Humphrey, 40
Locke, John, 49–50, 103, 105, 157–58
Lockhart, J. G., 165
Logan, Peter Melville, 7
London Congress for Experimental
    Psychology, 186
Lonsdale, Henry, 71

Macaulay, Catherine, 47
machinery question, 108. *See also*
    industrialization
MacKenzie, James, 26
Mackenzie, William, 126
madness, 28, 149, 156–58, 160–62, 171, 179,
    182, 186, 188
magic lantern, 156–57
magnetism, 40, 127, 154–55, 174–75
Malebranche, Nicolas de, 173
Malthus, Harriet, 85–86
management, 5, 8, 13, 23–29, 34–37, 102,
    104, 127–28, 144–45, 153, 189–90, 192–93
Marcet, Jane, 47
Marshall, John, 183
Martineau, James, 35
masculinity: and managing nervous dis-
    order, 6, 17, 134, 145, 185, 188, 189; and
    sensibility, 20, 43–44, 145, 185; and the
    will, 20–21, 183; importance in natural
    philosophy, 42–48, 145, 183
materialism, 9, 12, 15, 31, 67, 109, 111, 125,
    143, 150, 153–54, 160–61, 163–64, 170–71,
    186–87
mathematics, 11, 36, 48, 89, 92, 152. *See also*
    probability, statistics
Mathison, W. C., 93
Maupertuis, Pierre-Louis de, 166
Maxwell, James Clerk, 156, 184; and
    masculinity, 21, 183; experimental

techniques of, 68, 95, 97, 186; on color vision, 68, 80, 88–93, 94, 100, 186, 196; philosophy of science of, 21, 47; poor health of, 4

Maxwell, John Clerk, 90

Maxwell, Katherine (Dewar) Clerk, 92–93

May, Charles, 83–84, 218n83

Meadows, A. J., 128

mechanical philosophy, 6, 155; challenges to, 64–65

Mechanics' Institutes, 35, 39, 41, 100

mental-moral philosophy: defined, 200n42

Mercato, Michael, 173

mesmerism, 138, 154–56, 169

Metcalf, John, 207n4

meteorology, 40

Mialet, Hélène, 25

migraine, 10, 101. See also hemiopsy

Mill, John Stuart, 21–22

Milner, Isaac, 4

mind-body dualism, 6, 11, 31, 34, 102, 141–43, 145, 154–55, 171, 187, 195–96

mineralogy: observational practices in, 18

Mint, 136

miracles, 147, 149, 160

Mitchell, Maria, 48

More, Henry, 151

Moyes, Henry, 207n4

Müller, Johannes, 16, 174, 179

Murray, J. A., 77

Murray, John, 165

muscae volitantes, 125–26

Napoleon Bonaparte, 190–91

nationalism, 38–39; impact on natural philosophy, 11, 27–28

natural history, 11, 195; importance of color in, 62

natural magic, 148, 165–70

natural philosophers: authority on nervous disorder, 16–17, 22–23, 29, 131, 133, 137–38, 141, 190; core group defined, 14; visualizing capabilities of, 182–83

natural philosophy: defined, 11–12; importance of color in, 62–65; importance of vision in, 17–19; transformation to modern science, 12, 194–97

natural theology, 60–61, 151–52, 154, 172, 187, 194–95, 197

nervous disorder: and hallucination, 162–63; as English malady, 6–7, 10, 25; as middle-class epidemic, 26; as scholars' malady, 25–26; frequent appearance of in scientific texts, 4, 16; gendering of, 3, 17, 44–45; impact on natural philosophy, 6, 13, 17. See also dyspepsia; hemiopsy; hypochondriasis; hysteria

nervous narratives, 7

nervous system: as mind-body conduit, 15; as model of scientific and social organization, 5–6; compared to telegraph, 9–10, 111, 190–93; importance within physiology, 15; philosophy of, 14–17; physiology of, 12, 14–17, 187, 191–92

New College (Manchester), 71

Newnham, William, 163

Newton, Isaac, 153–55; nervous disorder in, 27–28, 44; on afterimages, 105; on color vision, 62–63; on optics, 63, 64; on vision, 68, 117, 118, 124

Newtonianism, 104, 152, 194

Nicolai, Christoph Friedrich, 161–63, 175

nitrous oxide, 4, 175, 177

nonconformity, 14, 31. See also Christianity: Dissenting churches

North British Review, 42

objectivity, 145; attempts to achieve, 30–31, 36, 45, 56–57, 68, 79–80, 89, 99–100, 143, 187, 191–92, 196; challenges to, 17–19, 37–42, 48–51, 102, 185, 195, 197; gendering of, 47–48

ocular spectra, 105–106, 135. See also hemiopsy

opium, 177. See also laudanum

optics, 154–55, 157, 167; importance of color in, 62–65; physiological (see

optics *(continued)*
vision: physiology of). *See also* achromatism; camera lucida; camera obscura; color; instruments, scientific; photography; polarization; spectroscopy
Oreskes, Naomi, 145
overwork: as cause of nervous disorder, 4, 10, 17, 26, 28-29, 101, 112-14, 117, 119, 122, 123-24, 129, 132, 134, 136-38, 140, 142, 144, 175, 187
Oxford University, 42, 76

Paget, James, 183
Paley, William, 152, 172
Parry, Caleb Hillier, 140
Pascal, Blaise, 173
pathography, 6
Pearson, Karl, 179
Péclet, Eugène, 18
Penn, William, 74
perception, 20; distinguished from sensation, 15, 124-25
personal equation, 127. *See also* vision: subjectivity of
phenomenology, 32, 50, 156, 159
Phillips, John, 112
Philosophical Magazine, 16, 122, 128
photography, 36, 137, 144
phrenology, 20, 140, 143, 163; theory of color blindness, 66, 72
physics: emergence as a discipline, 194
physiology: emergence as a discipline, 194-96
Plateau, Joseph, 63
Playfair, Lyon, 110
poets: compared to philosophers, 25-26
polarization, 50, 67-68, 82, 95, 134
Pole, William: color blindness of, 56, 93, 214n146; on color blindness, 88, 93-100
Pond, John, 36, 127
Pope, Alexander, 174
Porter, Roy, 24
Porter, Ted, 36
Porterfield, William, 126

Powell, Baden, 28
precision, 11, 81, 86
Preliminary Discourse on the Study of Natural Philosophy, 65, 132-33
Prescott, George, 192-93
Priestley, Joseph, 12, 14, 28, 59, 67-69, 146, 153, 177
Pringle, John, 69
probability, 104
Prony, Gaspard de, 33
provincialism, 31, 55-100; and problem of communication, 48-51, 191; and genius, 71, 89-90; and Quakers, 70-72, 74-79; as challenge to objectivity, 37-42, 79; blindness as metaphor for, 69-70; color blindness as metaphor for, 57-62
psychology: and halluncination, 170-72; emergence as a discipline, 170-72, 181, 194-96, 228n30
Punch, 59-60, 176
punctum caecum. See blind spot
Purkyne, Jan, 10, 28

Quakers, 150; in discussions of color blindness, 56, 69, 72, 74-79; understood as provincial, 70-72, 74-79

race: and superstition, 164-65; and visual acuity, 34, 59, 185
railroads, 86-88, 100
Ransome, Joseph, 72-73
reason, 20, 43, 48, 132, 171-72; and hallucination, 147, 149, 179; in Christian thought, 9, 147, 163-64
reflex action, 6
Reid, Thomas, 152-53, 158, 196
religion. *See* Christianity
Ricardo, David, 34-35
Rivers, W. H. R., 60
Robinson, Thomas Romney, 35-36
Roget, Peter Mark, 33-34, 177
Romanes, George, 183
Romanticism, 6, 13, 108-9, 146, 153, 157, 165, 169-70, 179
Royal Observatory, Greenwich, 14, 83,

145; observational practices of, 22, 35–36, 127–28
Royal Institution, 44
Royal Society of London, 14, 47, 50, 71, 78, 81–82, 87–88, 94–96, 140, 165, 182

Sabine, Edward, 40
San, 34, 44, 185
Sandras, Claude, 174
sanity. *See* madness
Saunderson, Nicholas, 207n4
Schumacher, Heinrich, 36
Schuster, Arthur, 184
scotoma, 101. *See also* hemiopsy
Scott, Walter: hallucinations of, 166; nervous disorder in, 165; on hallucination, 165–68, 170
Secord, Anne, 11, 37
Sedgwick, Adam, 77
self-help. *See* Smiles, Samuel
sensation, 22; distinguished from perception, 15, 124–25
sensibility, 20, 27, 42, 48, 145
sensorium commune, 67, 137
Sheepshanks, Richard, 81, 138
Showalter, Elaine, 3
Simms, William, 81
Simpson, James, 92–93
Smedley, Edward, 113
Smee, Alfred, 112
Smiles, Samuel, 119, 207n4
Smith, Adam, 33
Smith, Crosbie, 102
Smith, W. Anderson, 185
Smith, William, 26
Smyth, Charles Piazzi, 183
Society for Psychical Research, 186
Society for the Diffusion of Useful Knowledge, 41, 165
Society of Friends. *See* Quakers
Somerville, James, 117
Somerville, Martha, 47
Somerville, Mary, 45–47, 168–69, 188; nervous breakdown of, 4
South, James, 128

Southcott, Joanna, 160
Southey, Robert, 75, 108, 177
spectroscopy, 62, 84
Speke, John, 207n4
Spencer, Herbert, 192; nervous breakdown of, 4
spiritual autobiography, 6
spiritualism, 154–56, 169, 171, 186–87
Spithead Mutiny, 191
Sprat, Thomas, 50
Spurzheim, Johann Gaspar, 72
Stafford, Barbara, 45
state: involvement in the sciences, 12, 41, 77; natural philosophers' position within, 190, 196; power struggles with local governments, 38
statistics, 195; as method for studying color blindness, 80, 86–88; as method for studying hallucination, 181, 183–84, 186
steam engines: boiler explosions, 109–10, 145; compared to human body, 8–9, 102, 106–12, 142
Stewart, Dugald, 196; color blindness of, 56, 69–70; on division of labor, 33; philosophy of perception of, 27, 124–25, 158–59
Stokes, George Gabriel, 95–96
Struve, Otto, 127–28
subjectivity. *See* objectivity
supernatural, 123, 147–48, 151–53, 156, 159–61, 163, 166, 168, 170–72, 178, 185–86
superstition, 147, 149–51, 153, 156, 162, 164–65, 167, 169, 188
Swan, William, 91
Swift, Jonathan, 48–49
Syme, Patrick, 81
sympathy, 36, 140

Tait, Peter Guthrie, 111, 186
taste, 43–44; and color blindness, 50, 58–59, 69, 100
telegraph: compared to nervous system, 9–10, 111, 190–93; electric, 113, 190, 192–93; optical, 190–92, 227n3
Thackray, Arnold, 41

thermodynamics, 107, 109
Thomson, C. Poulett, 77
Thomson, Thomas, 78
Thomson, William (Lord Kelvin), 102, 113, 153
toleration, religious, 150–51
Torres Strait expedition, 60, 87
transient hemiopsia, 105. *See also* hemiopsy
transient teichopsia, 140. *See also* hemiopsy
Trevithick, Richard, 110
Trinity College, Dublin, 210n80
Troughton, Edward: color blindness of, 56, 81–83, 99
Tulley, Charles, 81
Tyndall, John, 86–87, 113, 143
Tyrrell, Frederick, 140

Unitarians, 41, 71
United States: scientific culture compared to Britain's, 11
Ure, Andrew, 107–8

van Helmont, Jean Baptiste, 173
Venus, transits of, 19
vision: compared to other senses, 18–19; importance in scientific methodology, 17–19; physiology of, 65, 122–23, 126, 168; subjectivity of, 17–19, 167–68. *See also* afterimages; blindness; blind spot; color blindness; color vision; eye; hallucination; hemiopsy; illusion, optical; nervous system; optics; perception; sensation
vision, color. *See* color vision
vitalism, 6
voluntarism, 149, 152–53, 190. *See also* deism

Wahrman, Dror, 38–39
Walker, Ezekiel, 157
Warburton, Henry, 119
Wartmann, Elie, 50, 59, 67
Watson, Richard, 4
Watt, James, 177; and steam engines, 110–11; poor health of, 4

Wedgwood, Josiah, 177
Werner, Abraham Gottlob, 68, 81
Wheatstone, Charles, 95, 190; hemiopsy of, 105, 140–41; on managing the natural philosopher's body, 28
Whewell, William, 35, 56, 76, 85, 133, 152; on Newton's sanity, 27–28; philosophy of science of, 12, 16, 37, 65, 143
Whisson, Stephen, 69
Whytt, Robert, 15, 196
Wilberforce, Samuel (Bishop of Oxford), 177
Wilkinson, J. Gardiner, 60
will: and hallucination, 170, 179; as masculine trait, 20–21, 48, 183; importance in natural philosophy, 6, 19–23, 20–22, 43, 104; its control over body, 16, 132; of God, 152–53
William IV, King, 76
Wilson, George: early death of, 112–13; on color blindness, 72, 78, 80, 86–88, 91, 100, 186; on Quakers, 78
Winnecke, F. A. T., 10
Wise, Norton, 102
witchcraft, 165
Wollaston, William Hyde, 78, 144–45; as heroic scientific figure, 119–22; autopsy of, 117; hemiopsy of, 103, 105, 115–22, 129–30, 140, 168–69
Wollstonecraft, Mary, 47
women: and embodiment, 23; and nervous disorder, 3, 44–45; place in natural philosophy, 31, 36, 44–48, 75, 90–91, 133–34, 145, 168–69, 185, 188, 189
Wood, G. W., 77
Wordsworth, William, 55
work, 101–2. *See also* efficiency; overwork
Wundt, Wilhelm, 80, 92
Wyvill, Christopher, 151

Yeo, Richard, 36
Young, Thomas, 62; on color blindness, 66–67, 72, 82; theory of color vision, 66–67, 82, 88–89, 91, 96, 98–99, 208n37